Circadian Networks

The Body's Distributed Timekeepers

Team Trex

Published by NobleTrex Press

For permissions and other inquiries, write to:

P.O. Box 3132, Framingham, MA 01701, USA

Contents

Introduction

In the early 1960s, a German physiologist named Jürgen Aschoff constructed a bunker in the Bavarian hillside. It was a peculiar facility, designed not to protect against falling bombs, but to exclude the passage of time. The bunker was soundproof, shielded from the electromagnetic fluctuations of the atmosphere, and, most importantly, completely cut off from the rising and setting of the sun. Aschoff invited volunteers to live in this isolation for weeks at a time, devoid of watches, radios, or any external cues that might whisper the time of day.

The scientific consensus at the time leaned toward the idea that biological rhythms were merely reactions to the environment-that we slept because it got dark and woke because it got light. If you removed the stimulus, the logic went, the rhythm should disintegrate into chaos.

But chaos is not what happened in the bunker. As the days blurred into weeks, the volunteers continued to wake, eat, metabolize, and sleep in a rhythmic cycle. They did not drift into random naps or erratic feeding frenzies. Instead, they maintained a "day" that was close to, but rarely exactly, 24 hours. Without the sun to reset them, their internal timelines began to drift, waking up a little later each day, yet the internal coherence of their physiology remained unbreakable. Aschoff had proven that the rhythm comes from within. We are not merely reactive creatures; we are predictive machines. We carry the rotation of the Earth inside our blood.

This book is an exploration of that internal rotation. It is a look into the invisible infrastructure that governs not just when you sleep, but how you heal, how you think, and how you age. For decades, biology treated time as a constant-a background variable that didn't need much attention. If you wanted to test a cholesterol drug, you gave it to a mouse; it didn't seem to matter if you gave it at dawn or at dusk. If you needed to measure a hormone, you drew blood whenever the patient arrived at the clinic.

We now know this approach is fundamentally flawed. We have discovered that we are not constant. We are oscillating beings. The creature you are at 8:00 AM is biologically distinct from the creature you become at 8:00 PM. Your insulin sensitivity, your immune response, your cognitive speed, and even the permeability of your gut lining fluctuate wildly across the solar day. To ignore time in biology is like trying to understand a musical melody by looking at a pile of sheet music without playing the notes. The components are there, but the meaning is in the timing.

The realization that life is time-bound has sparked a revolution in medicine and physiology, earning Jeffrey Hall, Michael Rosbash, and Michael Young the Nobel Prize in 2017 for unraveling the molecular mechanisms that make this ticking possible. But the story has expanded far beyond the genetics of the fruit fly. We have moved from studying a single "biological clock" to understanding *circadian networks*-a complex, distributed web of timekeepers that hums within virtually every cell of your body.

This is the central marvel we must confront: You do not have a single clock. You have trillions.

Inside the microscopic universe of your cells, a feedback loop of proteins is currently assembling and degrading, keeping time with the precision of a Swiss watch. This occurs in the neurons of your brain, but also in the cells of your liver, the muscle fibers of your calves, and the beta cells of your pancreas. These are the distributed timekeepers. When they are in harmony, the body functions with extraordinary efficiency. The heart anticipates the morning sprint; the stomach prepares for the midday meal; the immune system ramps up during the day when the risk of infection is highest and pivots to repair mode at night.

But when these clocks fall out of sync-a condition increasingly common in our electrified, 24/7 society-the biological discord can be deafening. We call this "chronodisruption," and its fingerprints are found on nearly every major disease of the modern world, from diabetes and depression to cancer and heart disease.

To understand why this system exists, we have to look back to the very origins of life. The most reliable fact of life on Earth, for the last four billion years, has been the rotation of the planet. Light, dark. Light, dark. This environmental predictability offered a massive evolutionary advantage to any organism that could internalize it. If a bacterium waits until the sun burns it to turn on its UV-protection genes, it is already too late. Survival favors the organism that *predicts* the sunrise.

This predictive capacity is what we call the circadian rhythm (from the Latin *circa*, meaning "around," and *dies*, meaning "day"). It is an act of biological clairvoyance. By looking at the molecular gears that drive this rhythm, we can see how an ancient feedback loop-one that likely evolved to protect DNA from solar radiation-was co-opted to orchestrate everything from the migration of monarch butterflies to the release of cortisol in a stressed executive.

However, the architecture of this system is not a simple hierarchy. For a long time, we viewed the Suprachiasmatic Nucleus (SCN)-a tiny cluster of neurons in the hypothalamus-as the supreme dictator of time. We thought it barked orders to the rest of the body, and the organs simply obeyed. The reality, as uncovered by researchers like Ueli Schibler, is far more democratic and far more interesting. While the SCN acts as the conductor, syncing the body to the sun, the organs possess their own autonomy. The liver, it turns out, cares less about the light in your eyes and more about the food in your stomach. The interplay between the conductor in the brain and the orchestra in the viscera is a delicate negotiation, one that maintains our internal alignment.

This negotiation is currently under siege. For the vast majority of human history, our light exposure was massive during the day and non-existent at night. Today, we spend our days in dim offices and our nights bathed in the blue glow of LEDs. We have detached ourselves from the solar signal. We eat at times when our metabolism is shutting down, and we force ourselves awake when our brain chemistry is pleading for sleep.

Till Roenneberg, a prominent chronobiologist, coined the term "social jetlag" to describe the chronic misalignment between our biological clocks and our social schedules. It is a form of torture we voluntarily inflict upon ourselves, living in one time zone during the workweek and attempting to fly to another on the weekend. The consequences are not just grogginess; they are metabolic catastrophes.

As we peel back the layers of this network, we encounter the primary inputs that set the system: light and food. The pathways that carry light from the retina to the brain are distinct from those that create vision; you can be completely blind and still have a functioning circadian clock, provided these specialized cells are intact. Understanding how light creates an electrochemical dawn in the brain is vital for grasping why checking your email at midnight is akin to drinking a double espresso.

Simultaneously, the timing of nutrition-a field now blossoming as "chrononutrition"-challenges the old dogma of "calories in, calories out." Research from labs like Satchin Panda's has shown that a high-fat diet consumed within a strictly defined time window results in significantly less weight gain than the same diet grazed upon throughout the day and night. It appears that the liver needs a fasting window to perform its nightly maintenance, a concept that is reshaping our advice on dieting and metabolic health.

But the reach of the circadian network extends into even more surprising territories. Consider the immune system. We often imagine our immune cells as sentinels that are always on high alert. In reality, immunity is energetically expensive. The body cannot afford to be at "Defcon 1" continuously. Instead, it schedules its defenses. Circadian regulation of the immune system ensures that white blood cells traffic to tissues likely to encounter pathogens during the active phase, while inflammatory responses are often suppressed. This is why your fever spikes in the evening and why the symptoms of rheumatoid arthritis are most agonizing in the morning-the body's natural anti-inflammatory suppression lifts just as you wake up.

This temporal gating has profound implications for how we treat disease, a field known as chronotherapy. If the target of a drug-a tumor cell, a cholesterol enzyme, a blood pressure receptor-is oscillating, then the efficacy of the drug must also oscillate. We are discovering that for some chemotherapies, the difference between a toxic dose and a curative dose is simply the time of day it is administered. The future of medicine may well involve prescriptions that specify "Take at 16:00" rather than the vague "Once a day."

Yet, as we push the boundaries of this science, we also face the inevitability of entropy. The circadian signal is robust in youth, but it frays with age. The amplitude of the wave flattens; the peaks become lower, the troughs shallower. This dampening of the rhythm is intimately linked to the fragmentation of sleep in the elderly and perhaps even the onset of neurodegenerative diseases. Is the loss of rhythm a symptom of aging, or a cause? And if it is a cause, can we tighten the mainspring of the clock to rejuvenate the body?

We are also entering an era of quantification. Wearable technology promises to move circadian science out of the dark bunkers and blood labs into the real world. We are on the cusp of being able to track our "internal time" as easily as we track our steps. But this data brings its own ethical complexities. If a workplace knows your chronotype-whether you are a "lark" or an "owl"-will they discriminate against you because you aren't biologically wired for the morning meeting? Or will we move toward a society that respects biological diversity, allowing schedules to bend to the individual rather than breaking the individual against the schedule?

The journey ahead is not just about dissecting anatomy; it is about changing your relationship with your own physiology. We often view our bodies as machines that should operate identically regardless of the hour, frustrated when we cannot focus at 3:00 PM or sleep at 3:00 AM. Understanding the circadian network invites a more compassionate and strategic approach to living. It suggests that there is a tide within us, and swimming with it is infinitely easier than swimming against it.

In the chapters that follow, we will dismantle the clockwork. We will witness the elegant molecular loops that define a biological second. We will trace the neural highways that carry the news of the sun to the depths of the organs. We will examine the physiological cost of the night shift and the potential of timing to enhance the potency of our medicines.

But before we examine the gears, we must appreciate the music. We are rhythmic creatures in a rhythmic universe. The ticking is not just a sound; it is the very pulse of life. Let us listen to it.

Chapter 1

The Invisible Metronome

Deep within the machinery of life, a chemical metronome beats independently of the rising sun, orchestrating chaos into rhythm. This chapter dismantles the cellular clockwork to reveal how a tangle of proteins can keep time with the precision of a quartz watch, tracing the scientific detective story that proved time is woven into our very DNA.

1.1 The Molecular Pendulum

A pendulum keeps time because it cannot help itself: gravity pulls it down, momentum carries it through, and the same forces that sustain its swing also limit it. The cellular clock works by a similarly satisfying logic. It is not a tiny stopwatch hidden in the brain and broadcast by some internal radio signal. It is a set of molecules that, by the very act of turning genes on, gradually create the conditions that turn those same genes off—and then, after a delay, remove the brake so the cycle can begin again. The result is a rhythm that most of our cells can generate on their own, even when cut off from the outside world.

At first glance, this sounds improbable. Genes are pieces of DNA; they do not *move*. Proteins are chemicals; they drift and collide. Where does anything like a smooth 24-hour swing come from? The answer is that biology is excellent at building delays. If you want a system to oscillate, you need at least three ingredients: a push, a pull, and a lag between them. The push is gene activation: one set of proteins encourages a set of genes to be transcribed into messenger RNA (mRNA). The pull is inhibition: another set of proteins eventually suppresses that activation. The lag is all the time-consuming steps in between: transcription, mRNA processing, export from the nucleus, translation into protein, chemical modification, assembly into complexes, and transport back into the nucleus. When these steps are chained together, the cell creates a loop that can take roughly a day to complete.

In mammals, the best-known version of this loop revolves around a pair of proteins called CLOCK and BMAL1. They are transcription factors, meaning they bind DNA and influence the rate at which particular genes are transcribed into mRNA. Together, CLOCK—BMAL1 act as activators: they sit on regulatory DNA elements near a family of "clock genes" and promote their transcription. Among the key targets are *Period* (*Per*) and *Cryptochrome* (*Cry*). The mRNAs produced from these genes leave the nucleus and are translated into PER and CRY proteins in the cytoplasm.

If the story ended there, it would be a simple ramp: more activation leads to more PER and CRY. The crucial twist is what PER and CRY do next. As PER and CRY proteins accumulate, they bind to each other and to additional partners, forming complexes that can enter the nucleus. Once inside, these complexes repress the activity of CLOCK—BMAL1, reducing the transcription of *Per* and *Cry* and many other clock-controlled genes. That closes the loop: the activators create repressors, and the repressors silence the activators. Over time, PER and CRY are modified by enzymes (notably kinases that add phosphate groups), tagged for destruction, and degraded. As repression fades, CLOCK—BMAL1 activity rises again, and a new cycle begins.

The beauty is that no one step has to be precisely 24 hours. The period emerges from the entire chain of events and from the way the parts tune each other. A few minutes saved in one step can be lost in another. The clock is less like a metronome with a single adjustable arm and more like a small orchestra that keeps tempo by listening and responding: each instrument is a biochemical process with its own speed, and the whole ensemble settles into a stable rhythm.

The most intuitive way to picture this is as a daily "argument" that takes time to unfold. CLOCK and BMAL1 open the floor by turning genes on. PER and CRY gradually build the case for shutting them down, but they cannot do it instantly, because they are not yet present in sufficient numbers, and because they must be assembled and escorted to the nucleus. Only after they arrive does the suppression become strong. Then, as PER and CRY are dismantled and cleared away, the suppression weakens, and the activators regain the microphone. A cell can run this loop on an ordinary Tuesday or in a petri dish under constant conditions. The loop is *self-sustaining* because it produces its own brake and then, by degrading that brake, releases itself.

A historical anecdote makes the logic feel less abstract. In the early decades of molecular circadian biology, researchers could see daily rhythms in behavior and physiology, but the missing link was mechanism: how could a gene "know" what time it was? Work in fruit flies helped convert a philosophical question into a biochemical one. By 1990, Paul Hardin, Jeffrey Hall, and colleagues showed that the mRNA of the *period* gene in *Drosophila* rises and falls across the day, hinting that gene expression itself cycles. That observation helped crystallize a feedback idea: if the protein product of a gene can influence the transcription of that same gene—directly or indirectly—then a loop can form. The conceptual leap is simple, but it changes the entire problem. A clock is no longer an unknown essence that genes must consult; it is something genes and proteins can *build.*

In mammals, the core loop is often described as a transcriptional—translational feedback loop (TTFL). The name is clunky, but it points to the key steps: transcription (DNA to mRNA) and translation (mRNA to protein) are both essential, and the feedback runs through the proteins produced. The TTFL language became prominent as the molecular parts list expanded and as experiments showed that the rhythm persists in isolated cells. Joseph Takahashi and others emphasized that circadian control is not confined to a handful of genes. CLOCK—BMAL1 and related factors drive rhythmic changes across the genome: waves of transcription, rhythmic recruitment of the transcriptional machinery, and changes in chromatin structure that make DNA more or less accessible at different times of day. The clock is not merely a timer attached to physiology; it is interwoven with the daily management of cellular labor.

Still, if CLOCK and BMAL1 simply turned on *Per* and *Cry*, and PER—CRY simply turned them off, why should the cycle be robust rather than collapsing into a steady state? In many feedback systems, negative feedback stabilizes: it holds a variable near a set point. A thermostat does not usually oscillate between tropical heat and winter cold; it hovers. Oscillations require the feedback to be delayed and, in a sense, overcorrecting. Biology creates that delay in several ways.

First, the repressors must *accumulate*. At the beginning of a cycle, there is little PER and CRY available, so the activators work relatively unhindered. Only after enough protein is made can repression become strong. Second, the repressors must *mature*. PER and CRY are not born ready to shut down transcription. They are modified after translation, they fold, they find partners, and they assemble into complexes. Third, they must *travel*. Much of their work occurs in the nucleus, but much of their production occurs in the cytoplasm.

Nuclear entry is not a trivial hop; it is regulated, and regulation consumes time. Fourth, they must *be removed.* Degradation is not instantaneous, and it is itself regulated by chemical tags and by the availability of the cellular machinery that dismantles proteins. Collectively, these steps make repression rise and fall with a smoothness that feels almost mechanical.

This is the heart of the pendulum metaphor: not an object swinging in space, but a molecular population rising, acting, and being cleared away. One can imagine PER and CRY as sand slowly filling the upper bulb of an hourglass. When enough sand has accumulated, it tips the mechanism and blocks the activators. Then the sand is drained away by degradation until the block can no longer hold.

A clean way to visualize the loop is to treat it like a clock face in which each "hour" is a molecular event. The activators at the top initiate transcription; the resulting mRNA leads to protein accumulation; protein complexes enter the nucleus and inhibit transcription; degradation relieves inhibition; and the cycle restarts. The diagram below places these events in a circle and uses color to emphasize the push—pull dynamic.

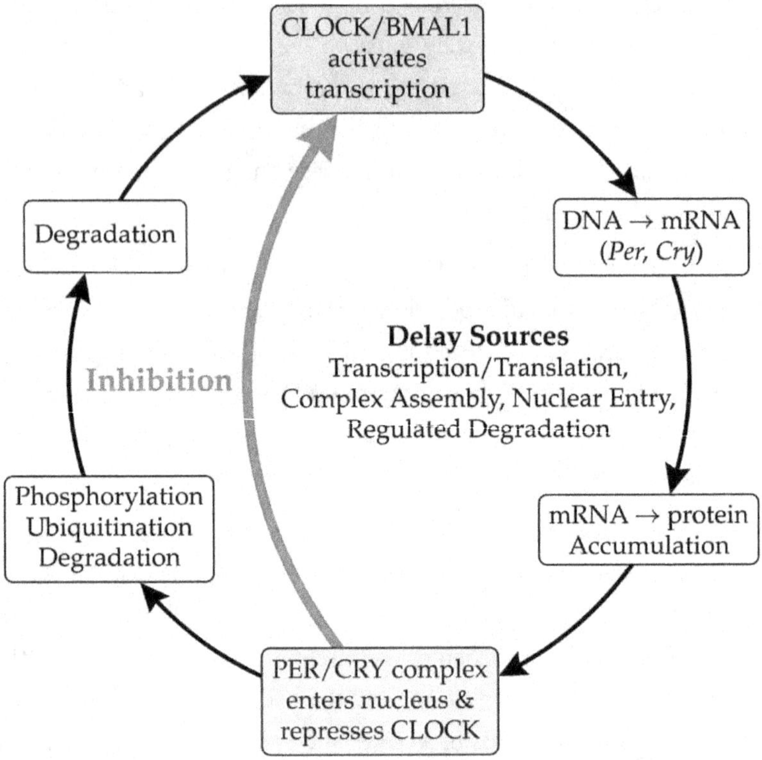

Seen this way, the clock is less mysterious, but it raises a deeper question: why is it so close to 24 hours? The Earth gives us a day, but cells do not have access to astronomical information. The period is an emergent property of reaction rates, binding affinities, and transport steps. Evolution tunes these parameters because organisms with internal cycles close to the environmental day gain an advantage: they can anticipate dawn and dusk rather than merely react to them. A clock that drifts too fast or too slow is harder to align with the world.

Tuning shows up vividly in genetics. In fruit flies, mutations in *period* can shorten, lengthen, or abolish rhythmic behavior, as Konopka and Benzer reported in 1971. In mammals, mutations in clock-related genes can shift sleep timing, alter hormone rhythms, and change the ease with which an organism adapts to shifting light schedules. The takeaway is not that one gene *is* the clock, but that the timing depends on many adjustable screws. Alter the stability of PER, change how quickly it enters the nucleus, or modify how strongly it represses CLOCK—BMAL1, and the cycle stretches or compresses.

One particularly revealing screw is phosphorylation: the addition of phosphate groups to proteins. Phosphorylation can act like a molecular Post-it note—“hold here,” “move there,” “degrade this”—except the note is an actual chemical modification that changes shape and binding behavior. PER proteins are heavily phosphorylated, and those phosphorylation patterns help control how long PER persists and when it is targeted for degradation. When PER is degraded faster, repression ends sooner, and the cycle shortens. When PER is stabilized, repression lasts longer, and the cycle lengthens. This is one reason circadian timing is not solely “in the genes” as static information; it is also in the life history of the proteins those genes encode.

There is another subtle reason the TTFL can produce stable oscillations: the loop is not a single line but a web of supporting feedbacks. In mammals, auxiliary regulators such as REV-ERB and ROR form interlocked loops that influence *Bmal1* expression and help sculpt the waveform—how sharp the peaks are, how deep the troughs go, and how resistant the rhythm is to noise. Carrie Partch and Carla Green highlighted in 2014 how these interlocked loops and post-translational controls shape the clock’s amplitude and phase. If the core loop is the pendulum, these additional loops are the escapement and gears: they do not create the basic swing, but they make it steadier, more precise, and better integrated with the rest of the cell.

The phrase "within virtually every cell" deserves to be taken literally. Liver cells, muscle cells, fat cells, immune cells, skin cells—many of them show circadian rhythms in gene expression when measured carefully. This is not just a curiosity. A liver cell does not need to be told, each morning, to start processing nutrients; it can anticipate. Immune cells can time aspects of surveillance and inflammatory readiness. Skin can coordinate DNA repair with the daily pattern of ultraviolet exposure. Even cell division shows daily patterning in some tissues. The clock is thus a way of scheduling incompatible tasks. Many cellular processes cannot be maximized simultaneously without conflict: growth competes with repair, energy storage competes with energy release. A 24-hour rhythm allows the cell to allocate resources in time as well as in space.

A concrete example makes the scheduling idea tangible. Consider the liver, which must decide, repeatedly, whether incoming calories should be burned immediately, stored as glycogen, or converted into fat. The molecular clock influences enzymes and transporters involved in these decisions, shaping daily rhythms in metabolism. If feeding happens at a time that systematically conflicts with the liver's internal program, the metabolic machinery can become misaligned: the enzymes expected to be high are low, and vice versa. Over weeks and months, this misalignment can contribute to metabolic strain. The pendulum matters because it helps coordinate *when* a cell is prepared to do particular kinds of chemistry.

It would be comforting if the TTFL were the only way to build a daily rhythm, because it is easy to narrate: genes make proteins, proteins regulate genes. Biology is not that tidy. Even within organisms that clearly use TTFLs, researchers have found rhythms that persist when transcription is inhibited, and there are clocks in other kingdoms of life that can run largely on post-translational chemistry. Cyanobacteria provide the cleanest demonstration. In work summarized by

Michael Rust and Susan Golden in 2011, a three-protein system (KaiA, KaiB, KaiC) plus ATP can generate circadian oscillations in a test tube: KaiC's phosphorylation state cycles with a roughly daily period. No DNA is required for the oscillation itself. This does not mean the mammalian TTFL is irrelevant; it means that nature has discovered multiple ways to make a day. The pendulum can be built out of gears (protein modification cycles) as well as out of a transcriptional loop, and in many organisms these layers interact.

That multiplicity makes an important philosophical point about the cellular clock. A clock is not defined by its parts list but by its behavior: it must oscillate with a stable period, resist being thrown off by small perturbations, and be adjustable by environmental cues. Different mechanisms can meet those requirements. The mammalian TTFL is one such mechanism, and it is deeply integrated with gene regulation, which is why it is so influential across physiology. But the presence of additional oscillatory processes helps explain why circadian biology is often robust: when one layer is weakened, another can keep some rhythm alive, like a building with both a staircase and an elevator.

If the core loop is self-sustaining, why does it not drift wildly from one day to the next? Part of the answer is that the clock is a *limit cycle*: once the system is oscillating, it tends to return to the same trajectory after small disturbances. In everyday terms, it has a preferred swing. A random fluctuation might make PER levels a little higher or lower at some moment, but the feedback structure nudges the system back toward its characteristic pattern. This is one reason circadian rhythms can be so reliable in the noisy environment of a cell, where molecules are constantly being made and degraded, buffeted by temperature changes, nutrient availability, and the ordinary chaos of life.

The other part of the answer is that the clock is not lonely. Within the body, cells communicate timing information through hormones, temperature rhythms, feeding cycles, and neural signals. Even within a tissue, cells can influence each other through local signals. But the striking thing about the molecular pendulum is that it does not require those signals to exist at all. Give a cell the genes, the proteins, the degradation machinery, and the ordinary chemistry of life, and it can produce a daily rhythm as naturally as it produces a membrane or copies its DNA.

There is a final, human-scale way to appreciate the elegance of the loop. A mechanical pendulum is a compromise with physics: it borrows regularity from gravity but pays for it with friction and the need for occasional energy input. The molecular pendulum is a compromise with chemistry: it borrows regularity from feedback and delay but pays for it with continual turnover. Proteins must be made and destroyed; mRNAs must rise and fall; complexes must assemble and disassemble. Timekeeping is not free. The cell spends energy to measure the day, not by consulting a calendar, but by building and dismantling the very molecules that mark its passing.

1.2 From Dark Cupboards to Fruit Flies

A daily rhythm can be dismissed in two easy ways. Call it a reflex—a body reacting to sunrise and sunset like a photo-sensitive toy. Or call it a habit—a learned routine, reinforced by mealtimes, clocks on the wall, and the social fact that most people sleep at night. Circadian biology had to defeat both dismissals before anyone could sensibly ask what molecules were involved. The crucial move was methodological rather than technological: take organisms away from the obvious signals of time and see whether the rhythm persists. The experiments were often simple enough to fit inside a cupboard, but the implications were unsettling. If a rhythm continues in darkness, in constant temperature, with no external schedule, then timekeeping must be *endogenous*—generated within the organism.

In 1729, the French astronomer Jean-Jacques d'Ortous de Mairan performed an experiment that reads, in hindsight, like a parable about scientific skepticism. He chose a plant that makes its own daily performance: the mimosa (often described as *Mimosa pudica* in popular retellings, though related plants show similar rhythmic leaf movements). During the day its leaves open; at night they fold. One could easily imagine that light directly forces the leaves open, and darkness forces them shut, like a hinge responding to heat.

De Mairan placed the plant in constant darkness and watched. The leaves continued to open and close with a roughly daily rhythm. The plant behaved as if it still knew when day should be, even when day never arrived. The demonstration was not yet a molecular explanation, but it was something more foundational: it made it respectable to believe in an internal clock at all. The mimosa was not *reacting* to the world in the moment; it was *predicting* the world based on its own internal dynamics.

The humble cupboard also makes a deeper point about biological time. A clock is usually judged by how well it matches the world, but the first question must be whether the clock exists in the absence of the world. De Mairan's plant was not perfectly punctual in darkness. The rhythm can drift, and the period can be slightly shorter or longer than 24 hours. That imperfection is a clue rather than a flaw: it suggests an oscillator that runs on its own and then, under normal conditions, is corrected daily by environmental cues.

It took centuries for this insight to become a systematic science, partly because living organisms are noisy clocks. Their periods vary with temperature, nutrition, age, and health. In the laboratory, researchers began to design increasingly careful "constant conditions" in which these confounds were reduced. Constant darkness or constant light, stable temperature, controlled feeding, and removal of social cues became the basic toolkit. What emerged was not a single rhythm but a family of rhythms: activity, body temperature, hormone secretion, sleep, urinary excretion, and many other measures show daily patterning that can persist even when the outside world is flattened into unchanging sameness.

A dramatic human example came from temporal isolation studies in the mid-twentieth century. Jürgen Aschoff and Ursula Gerecke (1967) reported observations from volunteers living for extended periods without time cues—no windows, no clocks, no radio announcements of the hour. In such conditions, many rhythms continue but "free-run," drifting away from a 24-hour day. Even more provocative, not all rhythms drift together. Body temperature might adopt one period, sleep another, and hormonal rhythms yet another, a phenomenon often described as internal desynchronization. This was not merely an experimental nuisance; it was evidence that biological timekeeping is distributed. The body does not behave like a single master clock with perfect control over every output. Multiple oscillators exist, and they can fall out of step when the usual synchronizing signals are removed.

For anyone who has felt jet lag, internal desynchronization is an oddly familiar experience. The mind may feel ready for sleep while the stomach seems ready for lunch, or alertness arrives at an inconvenient hour. The temporal isolation studies gave scientific legitimacy to that intuition. They also hinted at why molecular clocks might be found in many tissues rather than only in the brain: if the body contains multiple timekeeping units, then a purely central clock would have to do an enormous amount of micromanagement to keep every rhythm aligned.

Even with this conceptual foundation, the mechanistic question remained stubborn: *what* in an organism oscillates? Early answers were often physiological and vague: perhaps a metabolic cycle, perhaps a nervous signal, perhaps a hormone. The turning point came when the problem was reframed as genetics. If you could find an organism in which the period of the rhythm was altered by a single mutation, then the clock would no longer be a ghostly property of life. It would be anchored to something tangible and manipulable: a gene.

Fruit flies turned out to be ideal for this, not because they are especially rhythmic, but because they are especially amenable to genetic analysis. They breed quickly, can be kept in large numbers, and their behaviors can be monitored in a standardized way. In the early 1970s, Ronald Konopka, working in Seymour Benzer's laboratory, began screening flies for abnormal daily rhythms. The method was, in its own way, as simple and stubborn as de Mairan's cupboard. Choose a behavior that shows a daily pattern, keep conditions constant, and look for exceptions.

Two behaviors were particularly useful. One was adult emergence from the pupal case (eclosion), which tends to cluster at certain times of day. The other was locomotor activity, which can be recorded by placing individual flies in tubes where their movements interrupt an infrared beam. Under constant conditions, normal flies show a roughly daily pattern of activity. Konopka and Benzer (1971) identified mutants in which these rhythms were altered in striking and consistent ways. Some flies ran fast, with a period around 19 hours. Others ran slow, with a period around 28 to 29 hours. Some appeared arrhythmic, lacking a clear daily pattern.

The crucial observation was not only that the period changed. It was that *multiple* rhythms changed in the same animals. Eclosion and locomotor activity were both affected, implying that a shared underlying timekeeping mechanism influenced different outputs. This undermined the idea that each rhythmic behavior had its own independent timer. It also strengthened the idea of a central oscillator in the organism, even if its influence was distributed. Mapping experiments showed that the mutations affected a region on the X chromosome, pointing to a discrete genetic locus rather than a diffuse, non-genetic property of cells.

The gene became known as *period* (*per*). The name now feels inevitable, but at the time it was a confident claim: the gene seemed to alter the *period* of the organism's rhythms. The existence of *per* mutants had a psychological effect on the field. If one gene could swing a daily rhythm by hours, then the clock was not merely a metaphor. It was a mechanism with parts that could be broken, sped up, or slowed down.

Genetic evidence, however, does not automatically provide a molecular story. A mutation that changes a rhythm could, in principle, act indirectly: a sick fly might move differently, a stressed fly might eat differently, and these changes might masquerade as clock changes. The strength of the *per* story was that the mutations produced consistent, heritable alterations in the rhythm under constant conditions, and the same locus influenced distinct rhythmic outputs. Still, connecting the gene to a self-sustaining oscillation required a further insight: the gene's activity itself might cycle.

By the late 1980s and early 1990s, molecular techniques made it possible to measure mRNA and protein levels across time. Paul Hardin, Jeffrey Hall, and colleagues (1990) reported that *per* mRNA in flies cycles across the day. This was not merely another rhythmic readout. It suggested that the clock is built from feedback: the product of a gene somehow contributes to rhythmic changes in that gene's own expression. A gene that turns itself down after a delay is a plausible generator of oscillations. It is the same logic as the molecular pendulum: activation leads to accumulation, accumulation leads to repression, repression leads to decay, and decay permits activation again.

The elegance of the feedback idea is that it can produce a daily rhythm without requiring a daily environmental push. Light becomes an aligning signal rather than the engine. The oscillator can run in the dark, as de Mairan's plant did, because the oscillation is generated by the dynamics of gene regulation and protein turnover. The environment then corrects the phase, preventing drift and keeping internal time matched to external time.

It is worth pausing on how revolutionary this was as a way of thinking. In ordinary life, timekeeping is an act of comparison: you check your watch against a standard. In the body, there is no external standard available to each cell. A feedback oscillator solves this by being a process that can repeat itself with reasonable regularity, using only internal chemistry. Its "tick" is not a sound or a click; it is a sequence of molecular states.

The fly work also clarified something about the scale of biological explanation. A behavior like locomotor activity feels psychological, almost intentional. Yet the *per* mutants demonstrated that a single genetic change can reshape the timing of such behavior. That does not reduce behavior to a gene in any simplistic sense—the fly still needs muscles, neurons, sensory systems, and motivation-like states. But it shows that timing can be an organizing principle that sits above many systems, coordinating them without dictating their content. The clock does not tell the fly *what* to do; it biases *when* certain actions and physiological states are more likely.

In retrospect, the path from plants to flies reveals a set of experimental virtues that are easy to praise and hard to practice: patience, humility, and the willingness to build an argument from absence. It takes patience to watch leaf movements day after day. It takes humility to accept that an organism may be keeping time in a way that feels uncomfortably like intention. It takes methodological discipline to remove time cues and then trust what remains.

These experiments also anticipated a modern understanding of circadian systems as networks. De Mairan's mimosa was not only a proof of endogeneity; it was a hint that timekeeping is a property of living tissue, not merely of nervous systems. Aschoff and Gerecke's isolation studies suggested multiple coupled oscillators within one person. Konopka and Benzer's mutants suggested a genetic core that influences multiple outputs. None of these results require a single centralized clock in the naïve sense. Instead they point toward a hierarchy of coordination: internal oscillators that can run locally, coupled to one another, and corrected by environmental signals that are stronger and more reliable than any individual cell's chemistry.

A final anecdote brings the story back to the concrete experience of the laboratory. In many circadian experiments, the most important equipment is not a fancy machine but a locked door. A light leak can ruin weeks of data by providing an unintended cue. A researcher walking in with a flashlight can reset an animal's rhythms without realizing it. This fragility is itself a lesson: clocks are not only engines; they are *sensors*. They are designed to be adjustable. That adjustability is what makes them useful, but it also means that careful science sometimes begins with a surprisingly low-tech imperative: keep the organism in the dark, and keep the world from whispering the time.

Once the idea of an endogenous, genetically grounded oscillator was established, biology could begin to ask richer questions. Why are some rhythms easier to disrupt than others? Why do some tissues drift out of alignment under certain conditions? Why does a mutation change one person's sleep timing but not another's? Those questions belong to a world in which the clock is no longer a poetic metaphor but a testable mechanism, discovered in cupboards, bunkers, and tubes of quietly pacing fruit flies.

1.3 The Geometry of Rhythm

A rhythm is easy to mistake for a rule. If you wake at 7 a.m. for a week, it can feel as though your body has a schedule engraved into it. But biology rarely works by fixed appointments. It works by tendencies: pressures that rise and fall, gates that open and close, systems that become more or less willing to do something. The language of circadian science tries to capture those tendencies with a small set of geometric ideas borrowed from the mathematics of oscillations. The payoff is practical. Once the vocabulary is in place, it becomes possible to say why one person adjusts quickly to a time-zone shift while another feels wrecked for days, why a late-night bright screen can have a larger effect at one hour than another, and why some rhythms become fragile in illness even when the clock genes remain intact.

The most basic term is *period*. The period is the length of one full cycle. For a perfect clock it would be exactly 24 hours. In real organisms, under constant conditions, the clock often runs a little fast or a little slow. A human might free-run with a period slightly longer than 24 hours; a fruit fly mutant might free-run at 19 hours or 28 hours; a cultured cell line might drift depending on temperature and metabolic state. The period is not merely a number to record in a lab notebook. It determines how hard the environment must work to keep you aligned. A clock that naturally runs at 24.2 hours requires daily correction in the opposite direction, like a wristwatch that loses a bit of time every day and needs a frequent nudge.

If period is the cycle length, *phase* is where you are in the cycle. Phase answers the question, "How far along is the rhythm right now?"

It is often defined by a reference point—the peak of a hormone, the minimum of body temperature, the onset of melatonin secretion, the time of maximal activity in a fly. The same period can support many phases. Two people can both have 24-hour rhythms but one peaks earlier and one later. In everyday language, phase is close to "chronotype": whether you tend to be an early lark or a night owl. In laboratories, phase is measured with more care, but the intuition holds: phase is a position on the clock face.

The third term, *amplitude*, describes the strength of the rhythm. If you plot a rhythmic signal over time, amplitude is the height of the peaks and the depth of the troughs relative to the average. A high-amplitude rhythm has a clear, forceful day—night contrast. A low-amplitude rhythm looks flat and indecisive. Importantly, amplitude is not the same as "healthy" in every context, but many robust physiological rhythms do show strong amplitude: body temperature rises and falls with a noticeable swing, hormones show clear daily pulses, and sleep propensity changes sharply across the day. When amplitude collapses, timing becomes harder to read, both for the scientist with an instrument and for the body's own tissues that rely on rhythmic signals as cues.

These three properties show up in daily life in a way that can be surprisingly concrete. Imagine two people who both claim to be "fine on six hours of sleep." One goes to bed at midnight and wakes at 6 a.m. feeling alert; the other goes to bed at midnight and wakes at 6 a.m. feeling ill and foggy. Period is not the likely difference: both are living in a 24-hour world. Phase may be the difference: if the second person's internal night is shifted later, 6 a.m. may be closer to their biological midnight, a time when alertness systems are low and sleep drive is still high. Amplitude may also matter: if the second person's rhythms are dampened—for example by irregular light exposure, stress, or illness—the signals that normally promote consolidated sleep at night and consolidated alertness by day may be weak, leading to a smeared, unsatisfying compromise.

Circadian biology becomes more intuitive when rhythms are treated not as lines on a plot but as trajectories in a space of possibilities. This is where the word *geometry* earns its keep. Arthur Winfree, in *The Geometry of Biological Time* (2001), championed the idea that an oscillator can be pictured as a point moving around a closed loop in an abstract space. The loop is called a *limit cycle*. It represents a stable repeating pattern: even if the system is nudged off the loop by a disturbance, it tends to return and continue cycling. A pendulum pushed slightly does not usually stop swinging; it settles back into its regular arc. A biological clock behaves similarly, although the "arc" is a changing pattern of gene expression, electrical activity, or hormone secretion.

In this geometric picture, *phase* is simply the position along the loop. You can label one point as "dawn-like," another as "dusk-like," and so on, not because the system has seen the sun but because the internal state at those points resembles what it tends to be at those times under ordinary life. *Amplitude* becomes the distance of the loop from the center. A large loop corresponds to a strong oscillation: the system travels far from its average state and back again. A small loop corresponds to a weak oscillation: the system barely moves.

The most useful consequence of this picture is that it makes sense of *phase shifting*. When an external cue—most famously light—arrives, it does not simply tell the clock "it is now 8 a.m." Instead, it perturbs the oscillator. Depending on when the perturbation hits, it can push the system forward or backward along its loop. A pulse of light late at night might delay the clock (pushing dawn later), while the same light early in the morning might advance it (pulling dawn earlier). This dependence on timing is captured by a *phase response curve*: a map that tells you, for each phase, whether a stimulus produces an advance, a delay, or little change.

This is also why the phrase "light shifts the clock" can be misleading. Light does not act on an abstract dial; it changes biology. In mammals, light detected by the retina influences a small structure in the brain called the suprachiasmatic nucleus (SCN), a paired cluster of neurons above the optic chiasm. Those neurons contain molecular clocks, but they also communicate with each other, forming a network. David Welsh and Joseph Takahashi (2010) emphasized that individual SCN neurons can oscillate on their own, yet the network coupling makes the collective rhythm more precise and more robust than its parts. Geometry helps here too: instead of one oscillator, imagine many oscillators, each with its own phase. The SCN's job is not merely to shift a single hand on a single clock; it is to keep a community of hands moving together.

A concrete example makes this network idea less abstract. When someone travels from New York to Paris, the external world shifts by six hours in an instant. The SCN must realign, but it does not do so as a perfectly rigid unit. Some neurons may shift faster than others, creating a temporary spread of phases within the SCN. The output of the SCN—the timing signal it sends to the rest of the body—then becomes blurred. That blur can feel like jet lag: sleepiness and alertness arriving at inappropriate times, appetite losing its usual schedule, mood becoming irritable. In geometric terms, the network is temporarily less coherent; its phase distribution is wide. As coupling pulls neurons back into synchrony, the distribution narrows and the signal becomes crisp again.

This is where amplitude becomes more than a descriptive term. If the oscillation's amplitude is strong, the system has a well-defined phase: the loop is large, the trajectory is clear, and perturbations tend to shift the system along the loop rather than knocking it into confusion. If amplitude is weak, the system can be pushed toward a kind of *phaseless* state. Winfree described circumstances in which a strong perturbation collapses the amplitude so much that the system's phase becomes hard to define, like a watch whose hands have fallen off. When the amplitude later recovers, the phase may effectively be reset, sometimes unpredictably. This is not merely a theoretical curiosity. It suggests why severe disruptions—sleep deprivation, illness, sudden changes in light exposure, or shift-work schedules—can feel qualitatively different from minor schedule changes. The system is not simply shifted; it is temporarily destabilized.

A small everyday illustration comes from the common experience of staying up far later than usual under bright light. Some people can "reset" surprisingly quickly after a weekend of late nights; others feel as if their internal timing has been scrambled. Part of the difference can be explained by phase: the same late-night light hits different people at different biological times. But amplitude matters too. Someone with strong rhythms—regular morning light, consistent meals, stable sleep timing—may have an SCN network that holds together under perturbation. Someone with already dampened rhythms may be closer to an amplitude collapse, so a weekend disruption produces a deeper loss of coherence.

Period, phase, and amplitude also help distinguish a true circadian rhythm from a random fluctuation. Randomness can produce apparent cycles, especially in short datasets. The hallmark of an oscillator is not merely repetition, but *stability under disturbance* and *predictable response to perturbation.* If you tap a pendulum, you can predict how its swing changes. If you tap a noisy signal, it does not respond in a lawful way. In circadian biology, a rhythm that persists in constant conditions, maintains an approximately consistent period, and shows a stereotyped phase response to light or other cues is far more likely to reflect an underlying clock than a coincidental pattern.

Mathematical models make this intuition sharper. Daniel Forger and Charles Peskin (2003) built detailed models of the mammalian circadian clock using mass-action kinetics, treating molecular interactions as chemical reactions with rates and binding constants rather than as vague delays. Such models are valuable not because they replace experiments, but because they force clarity about what could plausibly generate a 24-hour rhythm and how robust that rhythm is to parameter changes. In a model, one can ask: if PER degradation is slightly faster, does the period shorten by minutes or by hours? If transcription is reduced, does amplitude drop gradually or collapse? These are the kinds of questions that connect the molecular pendulum to the lived experience of fragile or resilient timing.

The geometric vocabulary also helps explain an apparent paradox: the body can be both precise and flexible. Precision comes from the stability of the limit cycle and, in the SCN, from network coupling that keeps many cellular clocks aligned. Flexibility comes from the fact that phase is adjustable. A clock that could not be shifted would be useless in a seasonal world where dawn changes its time across the year. The same machinery that permits entrainment—daily alignment to the environment—also permits maladaptive shifts when the environment is artificial. Electric light at night, indoor living during the day, and irregular sleep schedules are not just "bad habits"; they are stimuli delivered at phases that can predictably delay or advance internal rhythms.

A concrete laboratory anecdote brings out the practical meaning of phase response. Circadian researchers learn quickly that the most dangerous time to enter an animal room is subjective night, when a brief light exposure can produce a large phase shift. Entire experiments have been compromised by a momentary light leak—a door left ajar, a corridor light spilling under a threshold. The same light intensity during subjective day might have little effect. The animals are not being dramatic; their oscillators are simply more sensitive to perturbation at certain phases. The phase response curve is written into their biology.

The same principle shows up outside the laboratory in subtler form. Morning light tends to advance human circadian phase, making earlier sleep and wake times easier; evening light tends to delay it, making later sleep and wake times more likely. This is one reason adolescence often brings later bedtimes: a developmental shift in circadian timing can coincide with abundant evening light exposure from screens and social activity, pushing the system toward delay. Understanding this as a phase phenomenon changes the moral tone of the conversation. It is not simply a matter of willpower versus laziness; it is a predictable response of a biological oscillator to a stimulus delivered at a sensitive time.

Amplitude has its own set of social and medical echoes. Many conditions associated with modern life—chronic stress, irregular schedules, fragmented sleep, metabolic disease—are linked with dampened rhythms in physiology. A low-amplitude rhythm can make the internal day feel less structured: appetite cues become less reliable, energy levels fluctuate unpredictably, sleep becomes lighter and more easily disrupted. The person may describe this as feeling "off" without being able to name why. From the perspective of the circadian system, the internal signals that usually provide a strong distinction between day and night have lost contrast.

Even the simplest circadian measurements illustrate why amplitude matters. Body temperature, for example, has a daily rhythm of modest size, often less than a degree Celsius in healthy adults. That may sound trivial, yet small temperature swings can influence enzyme activity and gene expression across tissues. In a high-amplitude state, the body broadcasts a clear temporal cue to its cells. In a dampened state, tissues that use temperature as a timing signal receive a weaker message. The timing network becomes harder to coordinate, not because any component has failed catastrophically, but because the signal-to-noise ratio has dropped.

One can think of period, phase, and amplitude as the three coordinates that locate a rhythm in timekeeping space. Period tells how fast the internal day runs. Phase tells what the system currently considers "morning" or "night." Amplitude tells how confidently the system distinguishes those states. Together they explain why two people exposed to the same environment can experience it so differently. The environment may be identical, but the internal geometry of the oscillator is not.

The most striking feature of this geometry is that it is simultaneously abstract and deeply personal. A curve on a graph, a loop in a phase space, a distribution of phases across a neural network—these are mathematical objects. Yet they correspond to the felt ease of waking, the heaviness of late-night alertness, the strange hunger at an unfamiliar hour, the difference between sleeping like a stone and dozing fitfully. The body's invisible metronome is not just ticking; it is tracing a shape. When the shape is stable and well-formed, daily life has a kind of quiet predictability. When the shape is distorted—shifted, stretched, dampened, or scattered across a network that has lost synchrony—time becomes harder to inhabit, and the day can feel less like a rhythm and more like a series of surprises.

Chapter 2

The Conductor and the Orchestra

Imagine a symphony where every musician possesses their own metronome, yet all look to a single conductor to begin. This chapter delves into how the brain's suprachiasmatic nucleus broadcasts time to the liver, heart, and muscles, and investigates why these peripheral organs maintain the autonomy to play their own distinct rhythms.

2.1 The Solar Interface

Light is not merely something the brain *sees*. It is also something the body *uses*. Every morning, photons arrive with a kind of authority: they announce not only that the world is bright, but that the day has begun, that temperature will rise, that food may soon appear, that muscles will likely work, and that danger (in an animal's evolutionary past) might be more likely in the open. The puzzle is how that outside fact—brightness—becomes inside facts: a higher heart rate before you stand up, a surge of cortisol that loosens glucose from storage, a liver that changes which enzymes it prefers to make, an immune system that shifts its posture from repair to surveillance. The suprachiasmatic nucleus (SCN), a paired cluster of neurons tucked above the optic chiasm in the hypothalamus, is the small piece of tissue that turns light into a timetable. It is less a lamp and more a switchboard.

The first surprise is that the SCN does not need the kind of visual detail used for recognizing faces or reading. It does not care whether the light comes from a sunrise, a phone screen, or a hospital corridor at 3 a.m. It cares about *irradiance*: how much light is present, when it rises and falls, and (to some extent) what color it is. This is why circadian biology belongs as much to the physiology of the eye as to the physiology of sleep. The retina is not simply a camera. It is a layered neural tissue that already performs computations. Among its most consequential computations is an internal measure of ambient brightness, built not only from the familiar rods and cones but from a rarer population of retinal ganglion cells that are themselves light-sensitive.

Those cells are called intrinsically photosensitive retinal ganglion cells, or ipRGCs. "Intrinsically" matters: they respond to light directly using a photopigment called melanopsin, and they also receive input from rods and cones. One can think of them as a hybrid sensor: part meter, part integrator. Their job is not to draw a picture of the world, but to report the state of the sky. They respond slowly and persistently, well-suited to the gradual arithmetic of dawn and dusk. In a dim room they may be quiet; in bright light they fire steadily, sending a message that can be read by brain circuits that never ask, "What am I looking at?" but only, "How bright is it, and when did that change?"

From the retina, this light information travels along a dedicated route to the hypothalamus. Some axons leave the optic nerve not to join the main visual relay but to peel off toward the SCN. This route is the retinohypothalamic tract (RHT). Its existence is one of those anatomical facts that feels almost philosophical: the eye has a private line to the brain's clock. The axon terminals of the RHT release neurotransmitters (notably glutamate, along with modulators such as PACAP) onto SCN neurons. The SCN reads these chemical messages as time-stamps: light at a particular internal moment shifts the clock in a particular direction, nudging it earlier or later. That rule—light late in the subjective night tends to delay, light late in the night-to-morning transition tends to advance—is one reason a bright screen can feel harmless while you are using it, but expensive the next day.

A historical anecdote makes the SCN's special status vivid. In the early 1970s, a series of lesion experiments in rodents showed that destroying a tiny region of the anterior hypothalamus abolished the normal daily pattern of activity. Animals that previously ran on their wheels at night became temporally untethered: still capable of moving, eating, and sleeping, but no longer doing those things at reliable times. This was not a failure of motivation, nor a paralysis, but a loss of temporal organization. The anatomical target of those lesions turned out to be the SCN. That discovery had the force of a map being drawn: if you want to know where the brain keeps time, start here. Later work refined the picture: the SCN is not a single metronome but a network, with neurons that couple to each other so that the ensemble becomes robust, resistant to noise, and unusually stable against some cues that reset other tissues.

If the RHT is the private line into the conductor's podium, the harder question is how the conductor's gestures reach the orchestra. It is tempting to imagine a single master signal—a hormone released at dawn, perhaps, that washes through the blood and makes every cell reset its clock hands. The body is more subtle. The SCN influences other tissues through multiple channels at once: endocrine signals, autonomic nervous system outputs, rhythms in body temperature, and rhythms in behavior (sleep and wake, feeding and fasting, movement and rest). These channels overlap, sometimes redundantly, sometimes competitively. That redundancy is not wasted design; it is a form of reliability. When light is strong and nights are dark, these signals tend to agree. When life becomes modern—shift work, travel, late-night light, irregular meals—the signals can disagree, and that disagreement becomes biology rather than mere inconvenience.

The dispersion begins close to home. The SCN projects to nearby hypothalamic regions that act as integrators and relays, including zones that influence sleep-wake state, thermoregulation, and autonomic output. From there, descending pathways reach preganglionic neurons in the brainstem and spinal cord, which in turn control sympathetic and parasympathetic nerves. These are the "automatic" nerves that adjust heart rate, blood vessel tone, sweating, digestion, and adrenal activity without asking permission. In parallel, the SCN influences endocrine axes by shaping the timing of releasing factors and pituitary outputs, which then govern peripheral endocrine organs.

A concrete example: consider the adrenal glands, small caps perched on the kidneys. They produce glucocorticoids (cortisol in humans, corticosterone in many rodents), hormones that help mobilize energy, modulate immune activity, and affect mood and vigilance. In most people, cortisol peaks in the early morning, rising before waking and then gradually declining across the day. This pattern is not just a reaction to waking up; it is partly anticipatory. The SCN helps set the daily rhythm of adrenal output through a combination of hormonal signaling via the hypothalamic-pituitary-adrenal (HPA) axis and direct neural routes that tune adrenal sensitivity. The bloodstream, in other words, becomes a broadcast medium for time: a hormone pulse can act as a systemic "good morning," reaching tissues that may never directly experience light.

Yet the SCN does not simply shout "morning" into the blood and call it a day. Different organs need different messages at different times. The heart is not a liver with valves. The liver is not a muscle with mitochondria under strain. The SCN's job is to provide a stable reference and to shape a landscape of cues so that tissues can align their local schedules without being forced into identical behavior. It is worth pausing on that distinction. Coordination is not uniformity. If all organs peaked at once, they would compete for resources and flood the blood with conflicting demands. A well-timed body is staggered: some processes ramp up as others settle down. The SCN sets the reference; each tissue interprets.

That interpretive quality can be seen in how peripheral clocks respond to different "time-givers," often called *zeitgebers*. Light is the dominant zeitgeber for the SCN, but feeding is a powerful zeitgeber for many peripheral tissues, especially metabolic organs like the liver. Glucocorticoids can shift many peripheral clocks as well. Temperature cycles—small daily swings within the normal physiological range—can reset peripheral tissues in culture, while the SCN network tends to resist such resetting, a kind of stubbornness that is useful: the conductor should not be easily pushed around by every fluctuation in the orchestra pit.

A practical, human-scale illustration often appears in jet lag. After a long flight across time zones, the eyes deliver local light signals to the SCN immediately; the SCN begins shifting. But the liver, gut, muscle, and adipose tissue also take timing cues from meals and activity. If you eat at the "wrong" local time because you are hungry on your old schedule, you effectively tell your metabolic organs a different story from the one your eyes tell your brain. The result is not merely sleepiness; it can be nausea, impaired glucose tolerance, and a strange mismatch between appetite and satiety. People often notice that the stomach "lags" behind the mind. That folk observation has a physiological basis: different clocks shift at different rates under different cues, and the SCN is only one part of the negotiation.

The retinohypothalamic tract is the beginning of the negotiation, and it is concrete enough to draw. The diagram below keeps the anatomy deliberately spare: a retina feeding the SCN, and then two illustrative output targets—the adrenal glands and the heart—reached by endocrine and autonomic routes. Many more targets exist, but these two capture the logic: one hormone-mediated, one nerve-mediated, both ultimately changing cellular gene expression and physiology.

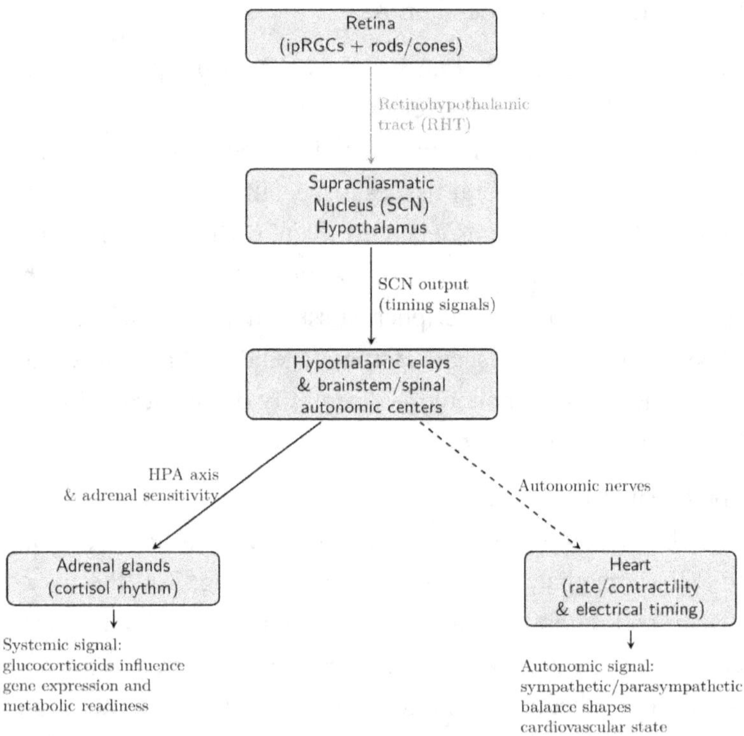

The SCN's output is often described as "synchronizing" the body, but synchronization can mean different things. A drummer can synchronize a band by giving a beat. A conductor synchronizes by shaping entrances, dynamics, and spacing. The SCN behaves more like the second. One reason is that the SCN's message is not a single waveform. It changes with day length. In winter, dawn and dusk arrive close together; in summer, they are far apart. Animals need to adjust physiology accordingly: when nights are long, activity, feeding, and metabolic processes may need different timing than when days are long. The SCN encodes aspects of photoperiod (day length) in the timing relationships among its own neurons and in the pattern of its outputs. Even for humans who live in temperature-controlled rooms under artificial light, seasonal shifts in mood and sleep hint that this encoding remains biologically meaningful.

To keep this grounded, consider a common morning experience: the "false start" of waking too early. If you wake at 4 a.m. after a short night and stare at the ceiling, you may feel simultaneously tired and oddly alert. Part of that tension comes from the fact that the circadian system and the sleep homeostat are not the same mechanism. The sleep homeostat measures prior wakefulness and pushes sleep when debt is high. The circadian system schedules wakefulness and sleep propensity across the day. In the early morning, the SCN-driven circadian program begins to prepare the body for daytime demands: cortisol rises, body temperature starts climbing, and autonomic balance shifts. If the sleep homeostat is still insisting on sleep, you feel the push-pull. Light at that hour is especially potent, because the SCN is biologically poised to interpret it as dawn. A bright bathroom light or a phone screen can become, unintentionally, an instruction.

The modern world provides countless opportunities for the SCN to receive ambiguous instructions. Electric light decouples brightness from solar time. Work schedules decouple activity from sunrise. Food availability decouples feeding from the day's energetic demands. The SCN is still doing what it evolved to do: interpret light as the most reliable cue and align internal time accordingly. The problem is not that the system is fragile; it is that it is faithful. A faithful translator will translate whatever it is given, even if the speaker is lying.

There is a poignant clinical angle to this fidelity. In intensive care units, patients can be exposed to constant dim light and frequent nighttime disturbances. They may be sedated, fed through tubes at odd hours, and deprived of strong day-night signals. Delirium, sleep fragmentation, and metabolic instability are common, and while the causes are multiple, circadian disruption is a plausible contributor. Some hospitals have experimented with lighting strategies that strengthen daytime brightness and protect darkness at night. The point is not that light is a cure, but that for the SCN, light is a language, and fluent speech matters when the body is stressed.

The SCN's translation from light to biology depends on the integrity of the eye-to-brain pathway. This is why circadian disruption can accompany certain forms of blindness. Not all blindness is the same for the circadian system. If rods and cones are lost but ipRGCs remain functional, circadian entrainment to light can persist even when visual perception is severely impaired. Conversely, if the pathway that conveys irradiance information is compromised, a person may have intact image-forming vision yet struggle with sleep timing. The separation of these functions is one of the most striking demonstrations that "light input" in biology is not a single channel. The eye contains multiple sensors serving different purposes, and the SCN listens to a particular one.

At the cellular level, the SCN's translation ultimately lands on gene expression. Light-triggered neurotransmission induces intracellular signaling cascades that affect transcription of core clock genes, shifting the phase of their feedback loops. The details can become technical quickly—proteins that inhibit their own transcription after a delay, cycles that take about a day, networks of coupled oscillators—but the intuitive point is simple: the SCN converts a physical event (photons) into a biochemical change (gene activity) that persists long after the light is gone. That persistence is what makes a clock rather than a reflex. A reflex ends when the stimulus ends. A clock uses the stimulus to reset an internal cycle.

The rest of the body eavesdrops on that cycle through the SCN's outputs. A morning cortisol rise changes how cells respond to inflammatory signals. Autonomic shifts change blood pressure and heart rhythm. Body temperature rhythms modulate enzyme kinetics across tissues. Sleep and wake reorganize the brain's metabolic housekeeping. Even posture matters: standing up in the morning is a predictable cardiovascular challenge, and the heart and blood vessels behave as though they have prepared for it. When the SCN is aligned to the environment, this preparation feels like normality. When alignment breaks, the day can feel like a series of small mismatches: hunger at odd times, fatigue at noon, alertness at midnight, a heart that races with coffee but will not settle into sleep.

The most revealing thing about the solar interface is that it is not mystical. It is wiring and chemistry. A minority population of retinal ganglion cells measures ambient light. Their axons form a tract that reports to a small hypothalamic nucleus. That nucleus, through overlapping neural and hormonal routes, biases the timing of physiology throughout the organism. The elegance lies in scale: a signal born in a sheet of tissue a fraction of a millimeter thick can reshape the behavior of organs meters away, not by commanding each cell directly, but by setting the conditions under which local clocks keep good time. Under a bright morning sky, the message is clear enough that almost every tissue can interpret it the same way. Under scattered, artificial nights, the message becomes a dialect, and the body does its best to translate.

2.2 The Autonomous Periphery

A wristwatch keeps time whether or not it is listened to. That is an awkward fact for any story that places the brain at the center of biological order, because it suggests a possibility that used to feel unlikely: perhaps organs keep time on their own. The SCN can broadcast a daily reference signal, but it does not follow that every heartbeat, every muscle contraction, every shift in fuel use is directly dictated from above. Much of the body behaves less like a set of puppets and more like a federation of skilled departments, each with its own internal timetable and its own ways of being convinced.

The first clue came not from a living animal, but from tissue that had been removed from one. If a clock is truly local, it should continue to tick outside the body—not indefinitely, perhaps, but long enough to show it is not merely echoing orders from the brain. For a long time, many physiologists assumed peripheral rhythms were weak ripples: present only because the brain and behavior imposed daily cycles of hormones, temperature, and meals. Then, in the early 2000s, the field gained a tool that made the invisible visible. Shin Yamazaki and colleagues (2000s) used mice engineered so that a core clock protein, PER2, was fused to luciferase, the enzyme that makes fireflies glow. In a dish, slices of liver, lung, and other tissues produced rhythmic pulses of light for days and days. The rhythm was not a transient afterimage; it persisted, with its own phase and its own quirks, like an instrument continuing to play when the conductor has left the room.

That image—organs glowing in a dish with a daily pulse—did more than delight the imagination. It rearranged the logic of the system. Peripheral tissues were not just passive targets of timing signals; they contained the molecular machinery of a circadian clock: feedback loops of gene expression in which proteins turn on the transcription of their own inhibitors, creating an oscillation with a roughly 24-

hour period. In many tissues, those oscillations are self-sustained for weeks under the right conditions. The SCN's job, then, is less to generate rhythm than to align rhythms so that the timing relationship between organs is consistent from day to day.

Alignment matters because the heart and skeletal muscles have different challenges from the SCN, and different incentives. The SCN sits in the brain, protected, rarely asked to perform sudden heavy work. The heart cannot choose to rest when the timing is inconvenient. It must accelerate for a sprint, recover from a fright, and maintain pressure in the face of posture changes. Skeletal muscle must shift between modes: quiet maintenance, burst power, endurance, repair. It would be inefficient to wait for a central command for every adjustment. A local clock allows an organ to *anticipate* predictable daily demands and to schedule its internal housekeeping accordingly.

Begin with the heart. Most people know that heart attacks and strokes are more common in the morning. The reasons are complex and not reducible to a single switch, but the pattern is hard to ignore: blood pressure rises toward waking, platelets become more reactive, and the cardiovascular system faces a daily "start-up" challenge as people go from lying down to standing, from fasting to feeding, from sleep to stress. It is tempting to treat these changes as a behavioral story: people wake up and begin doing things, so the heart responds. Yet the timing of many cardiovascular rhythms remains even when behavior is held relatively constant, and the heart's own cells contain clock genes that cycle.

A cardiomyocyte—a heart muscle cell—is not simply a pump component. It is a metabolic factory with strict quality control. It must manage calcium handling with exquisite precision, maintain ion gradients, repair damaged proteins, and decide which fuel to burn. A local clock provides a schedule for these decisions. Many of the genes under circadian control in the heart relate to energy metabolism, mitochondrial function, and electrical properties. The daily rhythm is not a decorative oscillation; it reshapes the baseline state of the organ. Even the heart's susceptibility to arrhythmias and its response to drugs can vary by time of day, implying that "the same dose" is not always the same physiological event.

Skeletal muscle offers an even more intuitive picture of why autonomy helps. Muscles are asked to perform at irregular times, depending on work, exercise, and emergency. Yet they also face a daily cycle of predictable states: feeding brings nutrients and insulin; fasting shifts the body toward mobilizing stored fuels; sleep reduces movement but increases repair and growth-related processes. Muscle tissue contains a large share of the body's mitochondria and glycogen stores, and it is a major site of glucose disposal after meals. A local clock can tune muscle insulin sensitivity, mitochondrial oxidative capacity, and the timing of protein synthesis and breakdown, shaping whether the muscle behaves more like a fuel sink or a fuel source.

Autonomy does not mean isolation. The heart and muscle are bathed in signals that carry information about time: temperature rhythms, sympathetic tone, glucocorticoids, melatonin, feeding-related metabolites. The important point is that these signals do not create clock machinery from scratch; they *set* and *adjust* clocks that already exist. This becomes obvious in experiments that change one cue while leaving another intact.

One of the most striking demonstrations uses feeding time. Francesca Damiola and colleagues (2000) showed that restricting feeding to an unusual time of day can shift the phase of peripheral clock gene expression, particularly in metabolic tissues like the liver, without shifting the SCN's rhythm. The SCN stays anchored to light, while the periphery begins to march to the timetable of meals. This is not a minor nudge. Under strong feeding schedules, some peripheral clocks can shift by many hours, even approaching a half-day inversion. If light tells the brain it is morning but food tells the liver it is "evening," internal time becomes a contested space.

Although the classic feeding experiments were centered on the liver, they reveal a general principle relevant to muscle: tissues do not all listen to the same cues with equal weight. A muscle that is repeatedly exercised at a consistent time can show shifts in local gene expression and performance characteristics, suggesting that activity is a time cue in its own right. In everyday life, this helps explain why a person can adapt to a habitual morning workout and then feel oddly weak when asked to lift heavy weights late at night, or why some athletes report predictable peaks in performance at certain times. The SCN can set the daily framework, but repeated local demands can sculpt local rhythms within it.

Glucocorticoids offer another clean separation between central and peripheral rules. Cortisol is often called a "stress hormone," but it is also a timing hormone: it rises with a circadian pattern even in the absence of acute stress. In 2000, Balsalobre and colleagues showed that glucocorticoid signaling can reset circadian gene expression in peripheral tissues and cultured cells. The SCN, by contrast, is comparatively resistant to being reset by glucocorticoids. This difference makes physiological sense. Cortisol surges during stress; if the SCN were easily reset by every stress response, the central clock would be jittery and unreliable. Peripheral organs, however, might benefit from hearing stress as a contextual cue: a signal that the day's plan has changed, that fuel should be mobilized and immune tone adjusted. In that view, cortisol is both a broadcast from the SCN and a messenger of acute circumstance, and peripheral clocks treat it as meaningful information.

The autonomy of peripheral clocks is also revealed by what happens when the SCN is removed or damaged. In animals, SCN lesions abolish coherent daily behavior and disrupt the normal coordination among organs. Yet the molecular clocks in peripheral tissues do not necessarily stop. They can continue oscillating, but they drift relative to one another, like musicians continuing to play without agreeing on where the downbeat falls. Over days, their phases can scatter across the 24-hour cycle. Within a single animal, the liver might peak at one time, the heart at another, muscle at another still. The body is not timeless; it is *unsynchronized.* That distinction is crucial for understanding why circadian disruption is not simply "loss of rhythm." Many rhythms persist. The trouble is in their relationships.

This is where the heart's autonomy becomes clinically relevant in an uncomfortably practical way. If the heart has its own clock, then being awake at night is not merely a matter of sleepiness. It means operating a cardiovascular system at a time when some aspects of its internal program may be set for rest and repair. Shift workers often describe a feeling of bodily resistance: an ability to stay awake but a sense of being "wrong-footed," with digestion, mood, and stamina failing to cooperate. Epidemiological studies have linked long-term shift work with higher risks of metabolic and cardiovascular disease, though causality in human populations is always hard to pin down cleanly because work schedules are tied to stress, socioeconomic factors, and lifestyle. Still, the biological plausibility is strong: internal misalignment makes ordinary demands more costly.

A concrete illustration can be found in an emergency night shift. Imagine a nurse who has worked several nights in a row. Her eyes see bright fluorescent light, her SCN is being tugged toward a nocturnal schedule, but she goes home in daylight and sleeps in a brightening world. She eats at odd hours because breaks are irregular. She drinks coffee at 2 a.m. because her job demands alertness, but coffee also shifts autonomic tone and can influence glucose handling. Her muscles are intermittently active—lifting patients, walking long corridors—but then she sits for charting. In such a week, the SCN receives a messy light signal; the periphery receives messy metabolic and activity signals. Even if her behavior is heroic and disciplined, her internal network may still be negotiating conflicting instructions. The consequences may show up not only as fatigue but as higher blood pressure, altered appetite, and impaired glucose tolerance at times that feel arbitrary but are, in fact, timed.

The autonomy of muscle clocks also matters for how we think about training and recovery. Muscle repair after exercise requires coordinated protein synthesis, mitochondrial biogenesis, and immune cell participation. Many of these processes are circadian-regulated. A local muscle clock can influence when inflammatory signals are most strongly interpreted as "repair" versus "threat," when satellite cells (muscle stem-like cells) are more ready to activate, and when mitochondria are primed to ramp up oxidative metabolism. This does not mean there is one best universal time to exercise. It means that timing is another variable, like intensity and nutrition, and that the muscle is not simply waiting for the brain to announce what time it is.

The laboratory evidence for local reset mechanisms is almost mischievously straightforward. In 1998, Balsalobre and colleagues showed that cultured cells, including fibroblasts, could be induced to display circadian gene expression rhythms after a "serum shock"—a sudden exposure to high serum concentration. The cells had no SCN, no retina, no behavior. Yet they responded to a systemic-like cue with multi-day oscillations. The implication is not that serum shock is a natural daily event, but that peripheral clock machinery is widespread and resettable by chemical signals that resemble what tissues experience in the body: hormones, nutrients, growth factors. Peripheral clocks are built to listen.

Listening creates a subtle problem: if peripheral clocks can be reset by local cues, how does the body avoid chaos? The answer appears to be coupling strength and cue hierarchy. Under ordinary conditions, the SCN provides a stable daily scaffold through light entrainment, and most systemic cues are themselves shaped by SCN-driven behavior and physiology. Feeding tends to occur during the active phase; temperature rises during the day and falls at night; cortisol peaks toward waking. Because these cues are usually aligned, peripheral clocks receive a coherent message. When one cue is artificially strengthened or shifted—a strict feeding schedule, repeated nighttime exercise, chronic stress with frequent cortisol surges—a peripheral clock may partially decouple and adopt a new compromise phase. That decoupling can be adaptive in the short term, but it can also create internal strain if different tissues compromise in different directions.

This is why the heart and muscles are such informative examples. They are not merely passive endpoints of metabolism; they are engines of action. The heart's clock must support reliable electrical timing and energy availability; muscle clocks must support both performance and repair. Both tissues face daily transitions that are predictable at the population level (most humans are diurnal, most activity clusters during daylight) but variable at the individual level. Autonomy gives them flexibility: a way to keep time even when central signals are imperfect, and a way to adjust to habitual local demands. The cost of autonomy is that it opens the possibility of disagreement. A single master clock would be easier to describe, but it would be brittle. A network of clocks is harder to coordinate, but it can adapt.

One can feel that networked design in an ordinary phenomenon: the second day of an early wake schedule. On day one, waking at 5 a.m. may feel like theft; on day two, the body often begins to anticipate. Hunger arrives earlier; bowel movements shift; the mind clears a little faster. Some of that is sleep debt and habit, but some of it reflects the ability of peripheral tissues to retime their programs in response to repeated cues: earlier meals, earlier movement, earlier light exposure. The retiming is not instantaneous. It is incremental, with different tissues adjusting at different rates. That slowness is often framed as inconvenience, but it is also stability. If the system reset completely after one late night, it would be constantly rewriting itself.

The idea of autonomous peripheral clocks also sharpens how one thinks about medical timing. If drug targets in the heart and muscle vary in expression or sensitivity across the day, then the effectiveness and side effects of medications might also vary. Some areas of chronotherapy—timing treatments to biological rhythms—have already entered practice in limited ways, but the broader promise is still uneven, in part because people's schedules and internal phases differ. Still, the principle stands: the same intervention can land in different physiological landscapes depending on circadian phase, and heart and muscle are among the landscapes that matter most.

Autonomy, in the end, does not make the SCN irrelevant. It makes the SCN's role more impressive. Coordinating independent oscillators is harder than commanding obedient ones. The heart and muscles keep time because they have to: the body cannot afford to rebuild its schedule from scratch each morning, nor can it wait for a central memo to decide when to mobilize fuel or repair proteins. They keep time locally, but they live in a body where the light-dark cycle still rules, and where coherence across tissues is the difference between a day that feels smooth and a day that feels like pushing against invisible resistance.

2.3 The Geometry of Timing

A well-timed body is not a chorus singing on a single note. It is closer to a score in which different sections enter at different moments, sometimes echoing one another, sometimes answering, sometimes leaving space. The difference matters. If every organ peaked at once—if alertness, digestion, immune activation, muscle power, and tissue repair all surged together—the result would be inefficient at best and unstable at worst. Biology rarely schedules like that. Instead it uses offsets, delays, and handoffs. The technical word is *phase*: where a rhythm sits within its cycle. The quieter truth is that health depends not only on having rhythms, but on having the right *relationships* among them.

Phase is easy to grasp when it is made physical. Picture two people walking around a circular track at the same speed. If they remain side by side, they are in phase. If one is consistently a quarter-lap ahead, they are out of phase by 90°. The track is the 24-hour day; the position on the track is internal time. Now imagine not two walkers but dozens, each representing a tissue or physiological process: the SCN, liver, heart, skeletal muscle, pancreas, adipose tissue, immune cells. They can share the same period (taking about a day to complete a lap) and still have very different phase relationships. The body's temporal organization is the pattern those relationships form. That pattern is the geometry of timing.

The SCN provides a reference. It is the best-anchored clock because light reaches it directly and because its neurons are tightly coupled, helping them maintain a coherent rhythm. Yet the SCN does not simply impose a single phase on the body. It sets constraints and provides signals that allow peripheral clocks to settle into stable offsets. Those offsets are not arbitrary. They reflect function.

Consider glucose handling after breakfast. The pancreas must be ready to secrete insulin; the intestine must be prepared for absorption; the liver must decide whether incoming glucose should be stored as glycogen or released into the blood; skeletal muscle must be receptive to insulin if it is to clear glucose efficiently. If these tissues were perfectly synchronized in a naive sense—all turning on the same genes at the same internal hour—they would not necessarily cooperate. Some processes need to anticipate others. Insulin sensitivity in muscle ideally rises *before* the biggest influx of nutrients. Hepatic glucose production should fall *as* feeding begins, not hours later. Digestion, absorption, storage, and oxidation are chained events, and chains require order. Phase lags are a way of writing order into physiology.

The language of phase also clarifies why circadian disruption often feels like a mismatch rather than a complete collapse. Many people with jet lag can still sleep, wake, eat, and work. The problem is that these actions land in the wrong internal context. Hunger arrives when the gut is sluggish. Alertness appears when the social world expects sleep. Muscles feel heavy at a time when, on the old schedule, they would have been primed. That sensation of wrongness is the lived experience of phase relationships that no longer align.

The geometry becomes visible in experiments that separate rhythms from one another. When the SCN is damaged in animals, daily behavior loses coherence. Yet peripheral tissues can continue oscillating, at least for a while, because their clocks are cell-autonomous. The striking feature is what happens *between* tissues. Instead of maintaining a reliable internal timetable, organs drift. A liver slice may peak in clock gene expression at one time, a lung slice at another, a muscle at another. The animal has not become arrhythmic so much as *decohered*. Coherence is a network property.

That insight has a mathematical flavor. In a network of oscillators, the state of the system is not fully described by the period of each node. It is described by the relative phases among nodes. Two clocks can keep perfect time individually and still be disastrously misaligned with each other. The body is exactly such a network. The SCN is a privileged node, but it is embedded in a web of couplings: hormones, autonomic nerves, temperature rhythms, feeding cycles, activity, and sleep. The couplings do not act like rigid rods; they act like springs. They pull, but they allow give.

Temperature is one of the most underappreciated springs. Body temperature in humans varies across the day by roughly 1°C, typically lowest in the late night and rising toward the day. That may sound trivial, but for enzymes and cellular processes, small temperature shifts can have meaningful effects. Peripheral tissues are sensitive to these oscillations. In culture, temperature pulses within a physiological range can reset peripheral circadian rhythms. The SCN, by contrast, tends to resist temperature entrainment, a robustness thought to arise from its network coupling. This creates an elegant division of labor. The SCN helps generate daily rhythms in behavior and physiology that produce a temperature cycle; the temperature cycle becomes a body-wide cue that peripheral tissues can read. It is a way of broadcasting time without sending a separate nerve to every cell.

This also explains why fever feels so disruptive. Fever is not merely discomfort. It is a forced shift in the body's thermal environment, and it can blur the clarity of the temperature rhythm that ordinarily helps align peripheral clocks. During illness, the immune system may *want* to disrupt normal schedules; energy is diverted, behavior changes, sleep becomes fragmented. The geometry of timing is not fixed; it is reconfigurable, and illness is one of the conditions under which reconfiguration is adaptive.

The hypothalamus plays an unsung role in shaping these relationships. The SCN projects into nearby hypothalamic regions that act as relay hubs, helping separate timing information into channels: sleep-wake regulation, thermoregulation, feeding behavior, autonomic output, endocrine rhythms. It is not necessary to memorize the names of these nuclei to appreciate the logic. A single central rhythm can be decomposed into multiple outputs with different timing, different strengths, and different sensitivity to context. That decomposition is how one clock can coordinate many functions without forcing them into lockstep.

A concrete example of competing couplings comes from restricted feeding. When feeding is limited to an unusual time of day in animals, peripheral clocks can shift, especially in metabolic tissues, while the SCN remains aligned to the light-dark cycle. The geometry changes in a structured way: the liver often shifts relatively quickly because it is directly engaged by nutrient flux and metabolic hormones, while other tissues can lag. The result is a temporary internal distortion—a set of phase angles that no longer match the usual pattern. If the feeding schedule is maintained, a new pattern may stabilize. If the schedule is irregular, the pattern may never settle. Either way, the key point is that the body's timing is not a single knob that turns all hands at once; it is a set of coupled hands that can be pulled in different directions.

This is where the notion of "staggered peaks" becomes more than a poetic phrase. Peak does not always mean maximum activity. It can mean the peak of gene expression for certain enzymes, the peak sensitivity of receptors, the peak propensity for a physiological state. Staggering peaks can reduce internal conflict. For instance, the body cannot maximize both tissue breakdown and tissue repair simultaneously without wasting energy. It cannot keep blood pressure at a daytime level throughout the night without imposing strain on vessels and heart. It cannot keep digestion in full swing during sleep without disturbing restorative processes. The circadian system handles these trade-offs by assigning phases: some processes are preferentially scheduled for the rest phase, others for the active phase, and many in between.

There is also a seasonal dimension. The SCN does not merely mark time within a day; it responds to day length. In many mammals, the SCN's internal organization changes with photoperiod. Neurons within the SCN can show phase differences across subregions, sometimes described as waves or spatial patterns of timing. Under long days, the distribution of phases across SCN neurons can broaden; under short days, it can compress. This internal geometry helps the organism encode whether it is summer or winter, influencing reproduction, metabolism, and behavior in species that depend on seasons. Humans are less overtly seasonal than many animals, but the machinery is still present, and subtle seasonal effects on sleep timing and mood suggest that the SCN retains some of this photoperiod sensitivity.

The idea that coupling can be attractive or repulsive, tightening or loosening synchrony, also provides a humane way to think about why some people feel more sensitive to schedule changes than others. A strongly coupled system resists perturbation. That is good for stability, but it can make adaptation slow. A more weakly coupled system adapts quickly but may be more vulnerable to internal drift under irregular cues. These are not moral traits. They are network properties, shaped by genetics, age, light exposure, social routines, and perhaps early-life experience.

Historical anecdotes in circadian biology often involve people forced to live without ordinary time cues. One of the most famous stories comes from isolation experiments in the mid-20th century, in which volunteers lived in caves or underground bunkers without clocks or sunlight, free to choose when to sleep and eat. Many drifted into schedules longer than 24 hours, revealing that the human circadian period is not exactly a day but close to it, and that light normally acts as a daily correction. What is less often emphasized is that such experiments also highlight internal geometry. When external cues are removed, sleep timing can drift, but so can appetite, body temperature rhythms, and performance. Sometimes these rhythms drift together; sometimes they separate. The separation is not a failure of willpower. It is a demonstration that different oscillators can couple strongly under ordinary conditions and uncouple when the environmental scaffold is removed.

Modern life creates a gentler version of that cave: constant access to light, food at all hours, irregular social schedules, and screens that deliver bright "dusk" at midnight. The body, however, continues to negotiate with the same rules: light anchors the SCN; feeding and activity tug peripheral clocks; temperature and hormones provide broadcast cues; sleep both follows and influences the timing network. When these cues agree, phase relationships settle into a stable geometry. When they conflict, geometry becomes warped.

Warping has recognizable signatures. One is the "social jet lag" experienced by many adolescents and working adults: late sleep on weekends, early waking on weekdays. The SCN tends to shift only gradually with weekend light exposure, while peripheral cues like meal timing can shift abruptly. Late-night eating on weekends can delay metabolic rhythms even if the SCN remains partially anchored by weekday mornings. The result can be a Monday morning in which the alarm drags the brain into wakefulness while the body's metabolic network is still running a weekend program. People often describe this as feeling "hungover" without alcohol. It is not imaginary. It is phase misalignment.

Another signature appears in shift work. On a night shift, light exposure at night can push the SCN later, but many shift workers still see daylight on the commute home, which can pull the SCN back earlier.

Meals may occur at 3 a.m. because that is when the break comes, but family dinner may still happen at 6 p.m. on days off. Exercise may be squeezed in whenever possible. Under such conditions, different tissues receive different time cues with inconsistent periodicity. The network may fail to settle into a single stable geometry. This chronic "never quite aligned" state is a plausible pathway by which circadian disruption becomes physiological risk rather than mere annoyance.

The geometry of timing also helps make sense of why some interventions work better than others. Bright morning light can be powerful because it provides a strong, unambiguous cue to the SCN, which then reshapes downstream signals. Regular meal timing can be powerful because it provides a strong, unambiguous cue to metabolic organs, which then reshape appetite, glucose handling, and energy levels. If a person tries to fix sleep by changing bedtime alone while leaving light exposure, meal timing, and activity irregular, the network is asked to do something difficult: realign phases without consistent couplings. Conversely, when multiple cues are aligned—morning light, daytime meals, evening dimness, consistent activity—the network receives a coherent message and tends to settle.

None of this implies that the body is a delicate clockwork that must be handled with tweezers. Humans have survived with erratic schedules for millennia: storms, newborns, feasts, famine, war, travel. The circadian network is resilient precisely because it is distributed and because it can reconfigure. The costs of misalignment tend to accumulate when misalignment is frequent, prolonged, or combined with other stressors like poor sleep quality, chronic stress, or metabolic overload. In that sense, the geometry of timing is less a fragile blueprint and more a dynamic truce among competing demands.

It is worth ending with a single, ordinary physiological handoff that captures the logic. In the late evening, core body temperature begins to fall, a cue that peripheral tissues interpret as night. Blood pressure tends to drop. The heart is still working, but the body shifts toward conservation and repair. Muscles reduce their readiness for sudden demand and increase programs involved in maintenance. In the early morning, before the eyes have even opened, the temperature rhythm begins to climb, cortisol rises, autonomic tone shifts, and the cardiovascular system prepares for standing and moving. These are not simultaneous events. They are staggered on purpose. A day begins not with a single switch being thrown, but with a sequence of timed adjustments that make waking feel, when things go well, like a smooth return to a body already prepared to be used.

Chapter 3

Chasing the Sun

We often think of our eyes merely as cameras capturing the world, but they serve a second, more ancient function as the body's primary time-setting interface. By tracing the specific photon pathways that synchronize our internal clocks to the solar day, we reveal how the invention of artificial light has fundamentally disrupted a billion-year-old biological pact.

3.1 The Blue Light Highway

Most people learn, early on, that the eye is a camera: light comes in, an image forms, and the brain interprets it. That story is true as far as it goes, but it leaves out a second job the eye performs with quiet efficiency. Alongside the circuitry that builds pictures, the retina runs a parallel service: a continuous audit of environmental light for the purpose of timekeeping. It is less interested in edges and faces than in a simpler question: *how bright is it, and what color is that brightness?* The answer is forwarded, not to the parts of the brain that help you recognize a friend across a street, but to a small hub in the hypothalamus that acts as the body's master clock.

The distinction matters because an "image" is a high-speed, high-resolution computation. The retina compresses the world into patterns of contrast, movement, and detail, feeding the visual system a stream that changes from moment to moment. Entrainment—the process of aligning internal time with the external day—needs something different. It needs a measurement that is slower, steadier, and more stubbornly statistical. Sunrise does not arrive as a single flash; twilight is not a sharp boundary. The signal that best reflects dawn or dusk is not a crisp picture but a gradual drift in the spectrum and intensity of ambient light. To keep good time, the brain needs a kind of light meter built into the eye.

This is where the retina's "non-image-forming" pathway comes in. The phrase sounds like an administrative category, but it captures a profound idea: light affects the brain even when it does not help you see. You can feel it when a bright morning makes you more alert before you have had any coffee, or when a luminous phone screen at midnight leaves you inexplicably awake. Those experiences are not merely psychological. They are, in part, the output of a dedicated neural route from retina to clock.

At the center of that route is a class of retinal cells whose existence, to many neuroscientists, once would have sounded redundant. The retina already had photoreceptors: rods for dim light and cones for color and detail. Why would it need more? The answer emerged gradually from a set of puzzles that refused to go away.

One puzzle came from the study of circadian rhythms in animals whose vision had been severely compromised. If rods and cones were the only way to detect light, then destroying them should have destroyed the ability to set the body clock by day and night. Yet many animals with profound degeneration of rods and cones could still synchronize their daily rhythms to a light-dark cycle. They could not see in any meaningful sense, but their physiology still "knew" whether it was day. The simplest explanation was that some residual rods or cones were surviving. The more interesting explanation was that the circadian system was listening to a different photoreceptor altogether.

A second puzzle came from human physiology. In laboratory studies of melatonin—a hormone secreted by the pineal gland at night that serves as one of the body's reliable markers of biological darkness—certain wavelengths of light suppressed melatonin more efficiently than others. The peak sensitivity did not match the standard curves used to describe brightness in vision. In other words, the light that "felt" brightest to the circadian system was not the same as the light that looked brightest to the conscious mind. That mismatch suggested a different biological sensor with different spectral preferences.

The sensor turned out to be a special subset of retinal ganglion cells. Ganglion cells are usually described as the retina's output neurons: they gather processed information from rods and cones (through intermediate bipolar and amacrine cells) and send it to the brain. For decades, the assumption was that ganglion cells were messengers, not detectors. Then, in the early 2000s, work led by David Berson and colleagues demonstrated that some ganglion cells are *intrinsically* photosensitive. These intrinsically photosensitive retinal ganglion cells are now commonly abbreviated as ipRGCs. The key point is embedded in the name: they do not merely relay signals from rods and cones; they can respond to light directly.

Their photopigment is melanopsin, a molecule tuned to short wavelengths—the blue part of the visible spectrum. "Blue light" is a phrase that is often used loosely in everyday conversation, but here it refers to a particular range: roughly the region around 460–480 nanometres, near melanopsin's strongest sensitivity. Blue photons carry more energy than red photons, but what makes them biologically potent for circadian timing is not mystical energy; it is simple receptor biochemistry. Melanopsin absorbs those wavelengths especially well, and ipRGCs translate that absorption into neural activity that informs the brain about environmental light.

ipRGCs also behave differently from rods and cones. Rods and cones are designed to capture rapid changes: they respond quickly, adapt quickly, and provide the raw material for sharp vision and motion detection. ipRGCs respond more slowly and integrate over longer time windows. That sluggishness is not a flaw. It is an engineering choice. A timekeeper benefits from averaging. If a cloud passes over the sun, vision needs to adjust without confusion; the clock, by contrast, should not treat a five-second shadow as a miniature nightfall. ipRGCs are well suited to measuring the "background" illumination that carries information about time of day.

Importantly, ipRGCs do not work in isolation. They sit within the same retinal network as the rods and cones and receive input from them too. The result is a hybrid sensor: intrinsically photosensitive, but also informed by the classic photoreceptors. This matters because it helps explain why the circadian system can respond to a broad range of lighting conditions. Under bright light, melanopsin's own contribution becomes especially strong; under dimmer conditions, rod input can still influence non-image-forming responses. The point is not that rods and cones are irrelevant to circadian timing, but that they are not the whole story.

The signal from ipRGCs leaves the eye by a route that looks, at first glance, like any other optic wiring. The axons of retinal ganglion cells bundle together to form the optic nerve. But within that bundle, different fibers are headed for different destinations. Many fibers travel to the lateral geniculate nucleus and onward to visual cortex, feeding image-forming vision. Another subset branches toward nuclei involved in reflexes and state regulation. The branch that matters most for timekeeping is the retinohypothalamic tract, a direct projection to the suprachiasmatic nucleus (SCN), a small paired structure above the optic chiasm in the anterior hypothalamus.

The SCN's importance in circadian biology is sometimes stated so often that it becomes a slogan: "the master clock." The more interesting truth is how such a small structure earns that title. The SCN is not a single metronome neuron but a network of thousands of cells, each with molecular oscillations of its own, coupled so that the network keeps stable time and sends coordinated signals to the rest of the brain and body. For that system to align with the world, it must be informed about light, and the retinohypothalamic tract is the most direct channel.

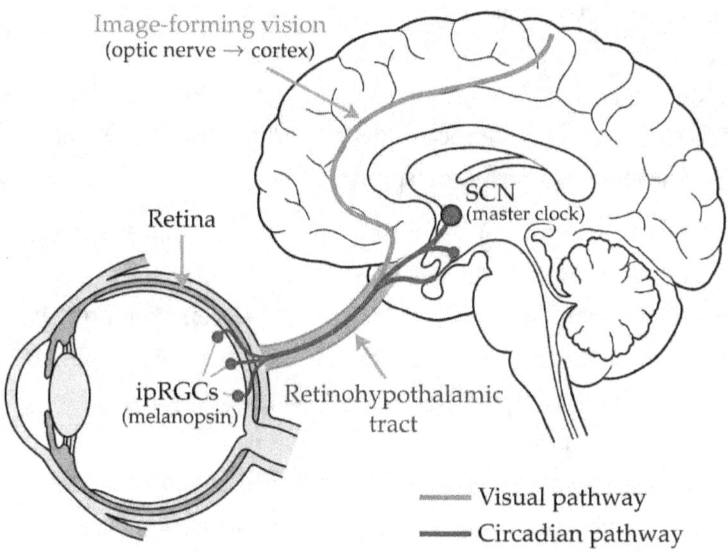

What travels along this "blue light highway" is not an image but a message about ambient illumination. That message is delivered in the language of synapses. The retinohypothalamic tract uses glutamate as a primary neurotransmitter, and also deploys a signaling peptide called PACAP (pituitary adenylate cyclase-activating polypeptide) in many fibers. When light activates ipRGCs, their terminals in the SCN release these chemical signals, nudging SCN neurons into a new phase. The molecular gears of the SCN can be sped up or slowed down in their daily cycle. Over repeated days, the clock aligns itself with the outside world.

The existence of this tract changes how one thinks about the eye. It is not merely a window but a sensor wired into the brain's control room. And it changes how one thinks about "seeing." A person who closes their eyes to sleep is not only blocking images; they are also, imperfectly, reducing the most powerful cue the SCN uses to decide whether it is day or night. The eyelids are not a perfect curtain. Enough light can pass through to stimulate ipRGCs, which is why sleeping in a brightly lit room can still perturb circadian signals, even if you never consciously register the light.

A concrete way to feel the difference between image-forming and non-image-forming pathways is to notice how little the circadian system cares about *what* you are looking at. If you spend the evening reading a paperback novel under a warm lamp, your eyes are certainly working: letters are resolved, pages are turned, stories are understood. Yet the circadian impact can be modest compared with reading the same text on a bright, blue-rich screen held at the same distance. The visual task is similar. The clock's input is not. The screen's spectrum and intensity, delivered directly to the retina, can drive ipRGC activity strongly enough to delay the internal night and suppress melatonin. Many people experience this as "I'm tired, but I'm not sleepy." The distinction is the body's division of labor: fatigue belongs to the homeostatic need for sleep that accumulates with time awake, while "sleepiness at the right time" depends heavily on circadian signals. Light can interfere with the latter even if the former is screaming.

The historical anecdote here is not a single dramatic eureka moment but an instructive shift in assumptions. For much of the twentieth century, researchers used the tools they had: measures of brightness designed for vision, like photopic lux, and models of retinal input that treated rods and cones as the only photoreceptors. Those tools worked well for predicting what people would report as "bright" and "dim." They worked less well for predicting how light would move circadian phase. The gradual accumulation of anomalies—melatonin suppression spectra that did not match visual sensitivity, blind animals that still entrained, reflexes and alertness changes triggered by light without corresponding changes in visual perception—forced a reclassification of what the eye does. The discovery of ipRGCs and melanopsin did not merely add a new receptor to the diagram; it legitimized a new category of light response: non-image-forming photoreception with its own optimization criteria.

Once you accept that category, everyday lighting begins to look different. Consider the simple act of stepping outdoors on a winter morning. Indoor light, even in a well-lit home or office, is often a few hundred lux. Daylight outdoors, even under cloud cover, can be an order of magnitude brighter, and on a clear day far more. To the visual system, the difference can feel like "somewhat brighter." To the circadian system, it can be the difference between a weak time cue and a decisive one. The ipRGCs are not impressed by your taste in décor; they are impressed by photons.

The retina's timekeeping circuit also explains why the color of light matters beyond aesthetics. Natural dawn and dusk are not only changes in intensity but changes in spectrum. As the sun sits low on the horizon, atmospheric scattering removes more short-wavelength light from the direct beam, leaving the warmer reds and oranges that painters and poets love. Indoors, the evening is often lit by sources whose spectrum does not drift toward red in the same way. Many "white" LEDs have a strong blue component, because blue light is used to excite phosphors that then emit broader wavelengths. The resulting light can look pleasantly neutral while still providing a robust melanopsin-weighted signal. A brain tuned by evolution to treat blue-rich light as "daytime" may receive, at 10 p.m., an accidental noon.

It is tempting to interpret this as a moral tale: nature good, technology bad. Biology is rarely that tidy. The same retinal circuit that makes evening light disruptive can be harnessed to help. Bright, blue-enriched light in the morning can advance circadian timing, making it easier for some people to fall asleep earlier and wake more easily. Carefully timed light is also part of clinical treatment for circadian rhythm sleep-wake disorders and for seasonal affective disorder, where morning light can relieve depressive symptoms in many individuals. The "highway" carries a message that can be helpful or harmful depending on timing, dose, and context.

The more unsettling implication is how automatic the system is. You do not have to believe in circadian biology for it to work. The retina does not negotiate with your intentions. It reports light to the SCN because it is built to do so. That is why a person can be sincerely committed to sleeping earlier and still sabotage themselves with an hour of bright screen time: the conscious plan lives in one set of brain circuits, while the clock-resetting machinery is quietly being driven by ipRGCs in another.

There is also something strangely comforting in the mechanism. The SCN is not guessing at time; it is being continuously calibrated by a reliable astronomical event. The planet turns. The spectrum shifts. The retina measures. A thin tract of neurons carries the report to a tiny nucleus, and from that nucleus the body coordinates countless rhythms: temperature, hormones, alertness, digestion, immune activity. The chain is delicate but not fragile. It is redundant, integrated, and adaptable. Even when the modern world tampers with the signal, the underlying logic remains the same: light is not merely something you see. It is something your nervous system *uses* to decide what kind of creature you are being at that moment—a daytime animal gathering energy and information, or a nighttime animal repairing itself in the dark.

3.2 The Geometry of Entrainment

The body's clocks do not simply *read* the day; they must continually *solve* it. The solution is necessary because an internal circadian rhythm is not a perfectly machined 24-hour device. Left to itself in constant conditions, the human clock usually runs a little long, drifting later by minutes each day. For a while that drift is subtle: you might not notice that your appetite, alertness, or sleepiness has slid a fraction of an hour. Over weeks, the mismatch would become unmistakable. If nothing corrected it, "midnight" inside the body would slowly detach from midnight on the clock on the wall.

The correction does not happen by consultation or willpower. It happens through geometry: the clock has a phase (where it is in its cycle), it has a period (how long the cycle takes), and it is being pushed by an external force—light—that arrives at different times relative to that phase. The same push can have opposite effects depending on when it lands. This is the crucial, slightly counterintuitive rule: a burst of light is not a generic "wake-up signal." Light is more like a hand on a rotating wheel. Push at one angle and the wheel speeds up; push at another and it slows down.

"Phase" is easiest to grasp with a familiar example. Imagine your daily rhythm as a circle marked with internal events: the rise of melatonin, the dip in core body temperature, the morning peak of cortisol, the times you naturally feel sleepy or alert. Wherever you are on that circle right now is your circadian phase. When researchers speak of advancing or delaying the clock, they mean shifting your position on that circle relative to the outside world. A phase advance makes you earlier: you get sleepy and wake up sooner. A phase delay makes you later: you get sleepy and wake up later.

Why does the clock need both advances and delays? Because the environment is not merely "day" and "night"; it is a repeating 24-hour pattern that your internal oscillator must lock onto despite its slight imperfection. If your clock's natural period is, say, 24.2 hours, it wants to drift later. To stay aligned with the sun, it needs a daily phase advance that pulls it back. If your clock's natural period were shorter than 24 hours, it would drift earlier and would need daily delays to compensate. In most humans, the tendency to run a little long makes the ability to advance especially important.

The body obtains those corrections by using light at the right times. Morning light usually produces advances; evening light tends to produce delays. Midday light, in many circumstances, does surprisingly little to the timing of the clock. That "does little" does not mean midday light is useless. It can influence alertness, mood, and other aspects of physiology. But for the narrow question of shifting circadian phase, there is often a kind of plateau: a region where extra light has a small timing effect compared with light delivered at the edges of the day.

This timing logic was mapped into a tool that has become one of chronobiology's most useful objects: the phase response curve, often abbreviated PRC. The idea is simple enough to describe at a dinner party, and deep enough to occupy careers. Take a person, measure their circadian phase, deliver a controlled pulse of light at a particular internal time, then measure how much their clock shifts. Repeat for many different timings of the same light pulse. Plot "time of exposure" against "resulting phase shift." The curve you get is the clock's instruction manual for light.

Human PRCs were not easy to obtain. To do them properly, you need to know internal time, not just clock time. That requires carefully controlled routines and physiological markers, such as the timing of melatonin rise in dim light. In the early 2000s, Khalsa, Jewett, Czeisler and colleagues produced a well-known human PRC to bright light using rigorous laboratory protocols. The curve captured what clinicians and shift workers had sensed intuitively for years: light is a powerful tool, but only when its timing respects the clock's own geometry.

A concrete example makes the logic vivid. Picture a nurse who has worked day shifts for years and sleeps from about 11 p.m. to 7 a.m. Her circadian clock expects morning light soon after waking and darkness in the late evening. Now she is assigned a week of night shifts. On the first night, the hospital is blazing with bright light at 2 a.m., a time her body treats as deep biological night. According to the PRC, light at that internal time is likely to delay the clock, pushing her rhythms later. That sounds helpful for adapting to nights. But then she drives home at 8 a.m. into bright sunlight, right at the part of the PRC that tends to advance the clock. The same day contains two strong pushes in opposite directions: a delay at night from workplace light, and an advance in the morning from sunlight. The result is not a clean adaptation but a tug-of-war that leaves the clock confused and the person exhausted. The nurse's suffering is not merely a matter of "bad sleep hygiene." It is a predictable consequence of applying light at the wrong angles on the wheel.

The shape of the PRC also explains a subtler frustration that many people recognize. Someone tries to "fix" a late schedule by going to bed earlier. They lie in the dark at 10 p.m., willing themselves to sleep. But their clock has not moved. Their internal night still begins later. They may fall asleep from sheer sleep pressure, but it is fragile, and they wake unrefreshed. Then they sleep in on the weekend, and the pattern entrenches. The missing ingredient is a properly timed phase advance, usually delivered by morning light (and often helped by avoiding bright light late at night). The geometry is not moral; it is mechanical.

A PRC is often drawn with "time of light exposure" on the horizontal axis, and "phase shift" on the vertical axis. Positive values are advances, negative values are delays. The curve dips below zero in the evening and early night (delays), rises above zero in the late night and morning (advances), and flattens around the middle of the day (a dead zone). The dead zone is not absolute. Its breadth depends on the intensity and duration of light, on recent light history, and on the state of the circadian system. But as a first approximation, it captures a practical fact: if you want to move your clock, the edges of the day are lever arms.

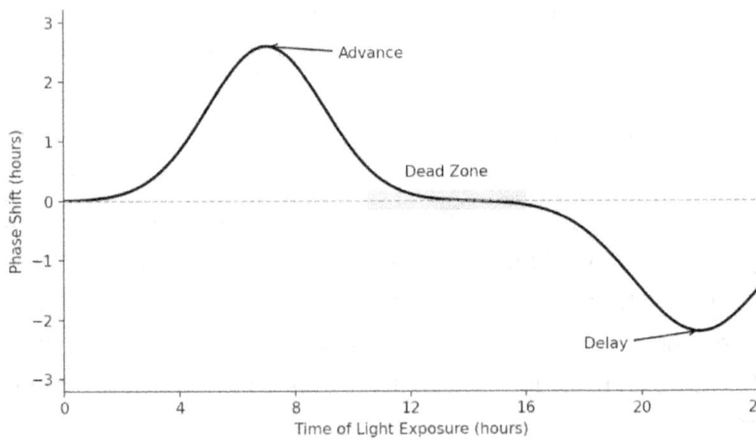

The PRC does not merely offer a picture; it implies a style of reasoning. It tells you that the question "Is this light good or bad?" is incomplete. The useful question is "Good or bad *when*?" A late-night light that delays the clock is unhelpful for someone trying to wake early, but it can be useful for someone adapting to a westward time-zone shift. A bright morning commute that advances the clock may be energizing for many people, but it can be counterproductive for someone trying to sleep later after a night shift. Chronobiology turns moral judgments about "screen time" into conditional statements about timing and goals.

There is another feature of the geometry that surprises people: the relationship between light intensity and circadian effect is not linear. Doubling the brightness does not necessarily double the phase shift or melatonin suppression. Many physiological responses to light show a saturating curve: dim light does little, moderate light does a lot, and then very bright light adds less than you might expect. The practical consequence is unsettling in a modern home. A person may imagine that only stadium-level brightness could possibly matter to the circadian system. Yet laboratory work by Zeitzer, Dijk and colleagues around 2000 demonstrated that the human circadian system is sensitive enough that ordinary indoor evening light can meaningfully suppress melatonin and shift circadian timing, especially when exposure is sustained.

That sensitivity makes evolutionary sense. In nature, twilight is not very bright, but it is reliable. A clock that ignored moderate light levels would miss dawn and dusk, the very signals that define the day's boundaries. The same sensitivity becomes a vulnerability when the environment supplies moderate light at the wrong time for long durations: living rooms lit until midnight, tablets used in bed, cities that glow through the curtains.

A further complication is that the clock's responsiveness depends on recent light history. Spend all day indoors under dim light, and the evening's lighting may have a stronger relative effect. Spend the day outdoors in bright daylight, and the same evening lighting may be less disruptive. This is not mystical resilience; it is adaptation. Photoreceptors, retinal circuits, and brain targets adjust their gain depending on the background. A chronobiologist would say that the system is context-dependent. An exhausted office worker would say, more simply, that a day without daylight makes the night feel electrically alive.

The PRC is often introduced as if it were a fixed curve, like the plot of a machine's specification. In reality, it is a family of curves. Change the duration of exposure, the spectrum (blue-rich versus warmer light), the intensity, or the person's prior light exposure, and the curve shifts in amplitude and sometimes shape. Even so, the qualitative rule holds with stubborn consistency: the timing of light relative to internal circadian phase determines direction.

One way researchers formalize this is with the mathematics of oscillators. A circadian clock can be treated as a limit-cycle oscillator: a system that, when perturbed, tends to return to a stable cycle. Light acts as a forcing term that nudges the oscillator forward or backward depending on its state. In this framing, the PRC is not an empirical curiosity; it is a signature of the oscillator's geometry in state space. That language can be intimidating, but the underlying intuition is familiar. A pendulum can be pushed to swing higher or damped down depending on when you apply the force. The clock is a biochemical pendulum distributed across cells and organs, and light is its main external shove.

The practical human world is full of unintentional experiments in this geometry. Jet lag is a textbook case. Flying east requires advancing the clock; flying west requires delaying it. People often do the opposite by accident: arriving after an eastward flight, they seek evening light to stay awake (which delays), and then sleep in the next morning (missing advancing light). The misery feels like a personal failing, but it is often a mismatch between the PRC and the traveler's choices. Athletes, diplomats, and flight crews now use light schedules precisely because the rules are so reliable. Even without special lamps, something as simple as stepping outside for a morning walk at the destination can act as a targeted phase signal.

The geometry also shows up in adolescence. Teenagers famously drift later, preferring later bedtimes and later mornings. Part of this appears to be biological: the circadian system shifts during puberty. Part of it is environmental: less morning light (because school and homes keep them indoors at dawn), and more evening light (from social life and screens). The PRC predicts that pattern will push the clock later. The consequences are not trivial. A teenager forced to wake at 6:30 a.m. for school may be waking during their biological night, when sleep is still strongly promoted and alertness is still suppressed. The result is not mere grumpiness; it is a chronic misalignment with measurable effects on mood, learning, and safety.

All of this can make the body clock seem like a tyrant, forever demanding that we live by solar time. Yet the geometry is also a source of flexibility. The clock can be moved, sometimes rapidly, when the light environment changes decisively. Field studies in which people exchange electric evenings for natural darkness, and dim indoor days for bright outdoor daylight, often find that circadian timing advances and aligns more closely with sunrise and sunset over surprisingly short timescales. The same plasticity that makes modern lighting disruptive also makes it possible to restore stronger time cues without changing one's genes or personality.

The most humane way to think about entrainment is not as obedience to nature but as a negotiation between an internal oscillator and a rotating planet. Light is the negotiator's language, and the PRC is its grammar. Morning light tends to say, "Start earlier." Evening light tends to say, "Start later." Midday light mostly says, "Stay steady."

When those messages are consistent day after day, the internal day and the external day can lock together with an elegance that feels effortless: sleepy at night, awake in the morning, hungry at the right times, mentally sharp when the world asks for it. When the messages are contradictory—a bright phone at midnight, a dim cave of an office at noon—the clock is still solving the geometry, but the problem has been written in a confusing hand.

3.3 The Artificial Day

A century ago, the night arrived with a kind of finality that is hard to recreate now. Lamps existed, of course, but they were dim, expensive, and local. The practical world narrowed after sunset. Work slowed, streets emptied, and the dominant environmental signal to the body was unambiguous: darkness had won.

The modern night is different. It is not simply that we can switch on a light; it is that we have made a parallel day that begins in the evening. Kitchens glow like laboratories. Phones and laptops keep a small rectangle of noon inches from the face. Cities cast a permanent orange-blue haze against the clouds, and even rural roads are stitched together by bright, high-color-temperature LEDs. The change seems cultural, even liberating—more time, more safety, more productivity. Physiologically, it is also a large-scale experiment in chronobiology: what happens when the retina's timekeeping circuitry is given a flattened day, with too little light when it expects "day" and too much when it expects "night"?

The circadian system is not fussy about lux meters or architectural taste. It is fussy about contrast. For most of human history, daytime was extraordinarily bright and nights were genuinely dark. Indoors, the contrast is often inverted: days are spent under a few hundred lux in offices and classrooms, while evenings extend under lamps, televisions, and screens that push light into the hours when biology is trying to declare night. The result is a smaller amplitude signal, like a radio station turned down and surrounded by static. The clock can still lock on, but it does so less decisively, and the consequences show up where people feel them most: in sleep timing and sleep quality.

Melatonin sits at the center of this story because it is one of the body's cleanest ways of announcing darkness. Produced by the pineal gland under control of the brain, melatonin rises in the evening, stays elevated through the night, and falls toward morning. It does not act like a sleeping pill in healthy people; rather, it helps coordinate internal night across tissues and nudges the body toward sleep-friendly physiology. When melatonin secretion is delayed, blunted, or interrupted by light, the body receives a mixed message: the clock may say it is night, but the retina is reporting conditions that are more like day.

This is where the modern spectrum matters. The retinal pathway that reports light to the master clock is unusually sensitive to short wavelengths, because melanopsin in intrinsically photosensitive retinal ganglion cells responds strongly in the blue range. Many contemporary light sources are rich in those wavelengths. A standard "cool white" LED often includes a pronounced blue component as part of the mechanism that produces white light. Screens, too, are typically blue-weighted compared with firelight or candlelight. The effect is not that blue light is uniquely evil; it is that blue-rich light is a particularly persuasive daytime signal to the circadian system.

A concrete example is familiar to almost anyone who has tried to be responsible at night. You turn the overhead lights down. The room feels dim and cozy. Then you open a laptop. The light seems subjectively small and manageable, but it is intense where it counts: concentrated, close to the eyes, and often blue-rich. You read, scroll, respond, plan. An hour passes. When you finally try to sleep, something is off. Your mind may be tired, but sleep feels oddly withheld, as if the body has not received official permission to switch modes.

Laboratory studies have made this intuition measurable. Evening light can suppress melatonin and delay circadian phase, and light-emitting screens can do so more than reading printed material under dimmer, warmer lighting. The point is not that screens have magical powers, but that they supply a tailored stimulus to the very retinal pathway that sets circadian timing: sufficient intensity, a spectrum that engages melanopsin, and long duration delivered at the most phase-sensitive part of the day. Under these conditions, the hormone that is meant to rise reliably each evening becomes negotiable.

The idea that ordinary evening light could have endocrine effects once sounded implausible. A historical clue to how quickly the environment changed comes from the early twentieth century, when electric lighting became widespread in wealthy cities. Diaries and social commentary from that era often celebrate the "conquest of darkness." Shops could stay open. Factories could run longer shifts. Streets could be lit for safety. Only later did scientists begin to appreciate that the conquest had a biological cost. In the same way that urban air pollution became visible only after industry scaled up, circadian light pollution became evident only after the evening grew reliably bright.

The indoors day is not just too bright at night; it is also too dim during the day. The circadian system uses daytime light not merely as a "wake up" cue but as a stabilizer that strengthens the difference between day and night. Bright daylight in the morning and afternoon helps anchor the clock, improves alertness, and can make the evening's melatonin rise more robust. A day spent under dim, unvarying indoor lighting can leave the clock more vulnerable to evening light, because the overall contrast has been reduced. This is one reason two people can experience the same evening screen habit differently: their daytime light exposure and routines change the background against which the evening stimulus is interpreted.

A useful way to think about the modern pattern is as a distortion of twilight. In nature, twilight is a brief, information-rich transition. Its spectrum shifts warmer, its intensity fades, and it arrives at a reliable time relative to solar events. Indoors, twilight is often stretched into hours, with a spectral composition that does not resemble sunset. The retina's clock-setting pathway reads that as a prolonged, ambiguous day. The result is often a subtle delay in circadian timing: internal night begins later, sleep onset becomes later, and waking at a fixed social time becomes harder.

This delayed timing connects to "social jetlag," a term used to describe the misalignment between internal biological time and the obligations of school and work. Many people live on two schedules: a workday schedule that forces early rising, and a weekend schedule that drifts later, closer to their preferred timing. The mismatch feels like jet lag without the plane: groggy Monday mornings, a sense of perpetual catch-up, and difficulty falling asleep on Sunday night. Evening light is not the only cause—genetics, age, and social habits matter—but it is an amplifier. A late chronotype exposed to bright evenings and dim mornings is pushed later still, increasing the mismatch with early schedules.

Outdoor artificial light at night adds another layer. Light does not respect property lines. Streetlights, billboards, and the general skyglow of a city can seep into bedrooms and shift the perception of night even with the eyes closed. Satellite observations have shown that artificial nighttime radiance expanded markedly in the early 2010s, driven in part by transitions to LEDs. The change is not only brightness but distribution: more places, more hours, and often a spectrum that is more effective for stimulating melanopsin than the older orange glow of low-pressure sodium lamps.

The physiological effects of outdoor light are difficult to isolate in everyday life because exposure is tangled up with many other urban variables: noise, stress, temperature, socioeconomic conditions. Yet the mechanism is straightforward. The same retinal circuitry that tells the master clock about sunrise is activated by a streetlamp. The same hormonal system that interprets darkness as permission for melatonin release can be suppressed by light leaking through curtains. The most disturbing aspect is that the effect can be subtle while still meaningful. A person may not wake fully or consciously register the intrusion, but the brain's light-sensitive circuits can still adjust the timing and depth of internal night.

This is one reason the public-health discussion around circadian disruption has become more pointed. Night shift work is an extreme form of living in an artificial day: bright light at biological night, sleep in daylight, meals at odd times, and social life misaligned with physiology. In 2019, the International Agency for Research on Cancer classified night shift work as probably carcinogenic to humans (Group 2A), citing mechanistic evidence linking circadian disruption to cancer-related pathways, while also acknowledging the challenges of exposure measurement and confounding in human studies. The classification is not a verdict that every nurse or factory worker will develop cancer, nor a claim that light alone is the culprit. It is a signal that disrupting circadian organization is biologically consequential enough to be taken seriously at population scale.

The same mechanistic logic helps explain why advice about lighting often feels contradictory. People hear "Get more daylight" and "Avoid blue light," and wonder how both can be true. The answer is timing and contrast. The circadian system wants bright, blue-rich light in the morning and daytime because that is the signature of the sky. It wants dim, warm light in the evening because that is the signature of dusk and fire. When technology supplies the opposite—

dim days indoors and bright blue evenings—the clock receives a scrambled version of the world.

The story becomes more personal when it intersects with the rituals of intimacy and rest. Consider the bedtime routine of a family with young children. The parent wants to wind down, but the only quiet time arrives after the household sleeps. That time often happens under a laptop glow, because email, entertainment, and the day's loose ends are all handled through a screen. The parent may not be chasing productivity; they may be chasing solitude. Yet the physiology is indifferent to motives. If the light delays melatonin and circadian phase night after night, sleep becomes later and shallower, mornings become harsher, and the parent's limited rest becomes more fragile. The emotional texture of modern life—always reachable, always illuminated—has a hormonal footprint.

There are design choices that can soften the artificial day without romanticizing the past. Lighting that changes across the day, brighter and cooler in the morning and dimmer and warmer in the evening, better matches the retinal system's expectations. Dimmer bedrooms, blackout curtains, and the elimination of small but persistent light sources (indicator LEDs, hallway spill) can restore a sense of night. Morning outdoor light, even for a short period, can provide a strong anchoring cue. Screen settings that reduce short-wavelength emission in the evening may help some people, though the total brightness and duration of exposure matter as much as spectrum. The point is not perfection; it is restoring a high-contrast signal: day that looks like day, night that looks like night.

It is also worth keeping humility about individual differences. Some people can read on a tablet at midnight and fall asleep as if nothing happened. Others are exquisitely sensitive. Age shifts sensitivity too: children often have clearer lenses and larger pupils that transmit more light to the retina; older adults may receive less short-wavelength light at the retina due to lens changes, yet still experience circadian disruption through other pathways. Chronotype, mood, medication, and prior light exposure all modify the response. The artificial day is a population-level phenomenon, but it lands in individual bodies with different vulnerabilities.

One of the most striking demonstrations of how quickly the system can respond comes from situations where modern lighting is temporarily removed. Camping trips provide a kind of accidental experiment. People who spend several days outdoors, with bright daylight and only dim firelight after sunset, often find that their sleep timing advances. They become sleepy earlier, wake closer to dawn, and report that the change feels effortless rather than forced. The effect is not a mystical return to nature; it is the circadian system receiving a clean, high-amplitude light-dark signal and doing what it evolved to do.

The deeper implication is that the artificial day is not merely a backdrop; it is an active ingredient in health and behavior. Light shapes hormone secretion, body temperature rhythms, and the timing of sleep propensity. It influences alertness in the moment and shifts the clock over days. It interacts with social schedules to produce chronic misalignment for many people. And because light is delivered through the eye's non-image-forming pathways automatically, it operates below the level of conscious control.

The world we have built is brightly legible, safe to navigate, and rich in possibility after sunset. It is also chronobiologically persuasive, whispering "day" late into the night and muting the daylight that should anchor our clocks. When darkness finally arrives—a lamp clicked off, a screen put away—the body does not instantly agree. Melatonin rises on its own schedule, the clock keeps its own counsel, and sleep comes most easily when the environment speaks the same temporal language as the brain.

Chapter 4

Dining with the Clock

While the brain sets its watch by the sun, the liver and gut set theirs by the arrival of nutrients. This second system of timekeeping allows metabolic organs to anticipate fuel, but it operates with a dangerous independence: when meal times drift late, these peripheral clocks uncouple from the brain, creating a state of internal jet lag that disrupts everything from blood sugar to immune defense.

4.1 The Metabolic Shadow Clock

Imagine two clocks in the same house. One hangs on the wall in the hallway: it keeps *social time*, resetting each morning when sunlight pours through the windows. The other sits quietly in the kitchen, keeping *meal time*, resetting not by dawn but by the clatter of plates, the first mouthful of breakfast, the slow fade of insulin after dinner. Most days, these clocks agree well enough that you hardly notice they are separate devices. But when meals drift into the hours your body associates with sleep, the kitchen clock refuses to be polite. It keeps time its own way, and the rest of your physiology has to live with the argument.

The hallway clock is the suprachiasmatic nucleus (SCN), a small cluster of neurons in the hypothalamus that takes its instructions from light detected by the retina. The SCN is often called the "master clock," but that title can mislead. It is less a single tyrant and more a conductor who keeps a beat for an orchestra that can, under the right conditions, play in several time signatures at once. The liver is one of the most assertive musicians in that orchestra. It can keep its own circadian rhythm—a roughly 24-hour cycle in gene activity and metabolism—and it can reset that rhythm based on when food arrives, even if the SCN is still faithfully counting light and dark.

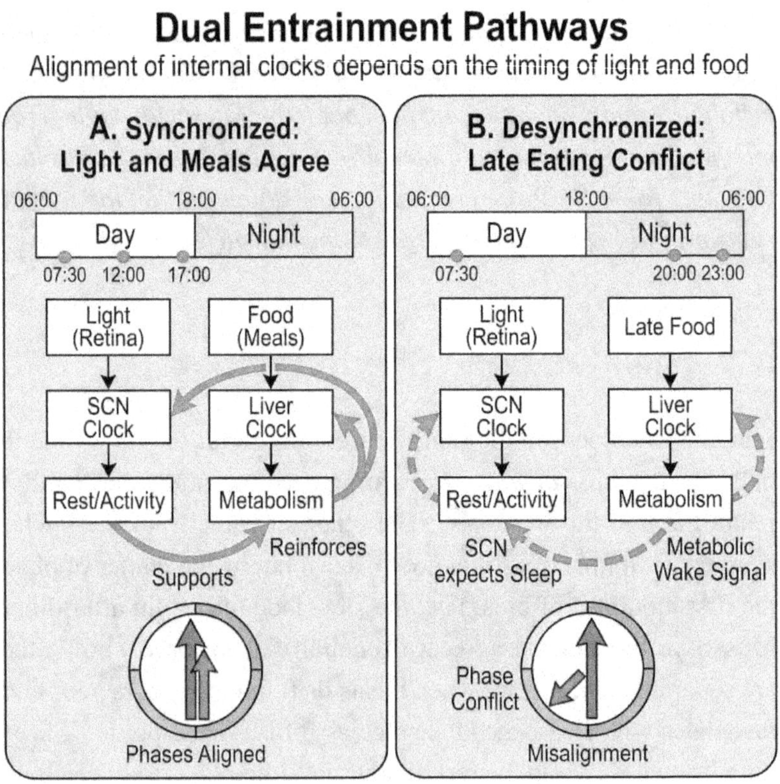

Calling this a "shadow clock" captures something psychologically familiar. You can go to bed at your usual hour, in a dark room, with your phone face-down and your sleep mask on. Your SCN receives the message: night has arrived; lower alertness; coordinate a hormonal landscape suited to rest. But if you eat a large meal late—or graze through the evening until you fall asleep—your liver receives a different message: nutrients are incoming; store energy; process glucose; package fats; handle a surge of amino acids. The body is not merely receiving two kinds of information. It is trying to run two kinds of *programs*.

The liver deserves this attention because it is a chemical switchboard. After a meal, it decides what to do with glucose: store it as glycogen, send it back into the blood, or convert it into fat when storage is full. During fasting, it reverses the flow: it breaks down glycogen, makes new glucose (a process called gluconeogenesis), and eventually shifts toward producing ketone bodies. It also detoxifies drugs and alcohol, manages bile production, and choreographs the traffic of lipids in and out of the bloodstream. These jobs cannot be done equally well at all hours. The evidence that they *aren't* is written into the daily rhythm of liver gene expression.

A circadian clock, in its simplest form, is a biological oscillator: a set of molecules that produces a self-sustaining rhythm with a period close to a day. In mammals, the core clockwork is built from feedback loops in gene regulation. Proteins such as CLOCK and BMAL1 help turn on the transcription of other clock genes (including *Per* and *Cry*), whose protein products later inhibit the activity of CLOCK and BMAL1, eventually lifting the inhibition and starting the cycle again. The details are intricate; the principle is easy to hold. A clock is a delayed negative feedback loop that keeps time by repeatedly building itself up and knocking itself back down. The SCN contains thousands of these oscillators coupled together, which makes its rhythm robust and relatively resistant to disruption. A liver cell contains its own version of this machinery, but liver cells are also exceptionally attentive to the metabolic state: to insulin, glucose, fatty acids, and the chemical cues that tell a cell whether it is in the fed state or the fasted state.

A historical turning point came from experiments that look simple on paper but were conceptually daring. In 2000, Francesca Damiola and colleagues reported that restricting feeding to an unusual time of day in rodents could shift the daily rhythm of clock gene expression in peripheral tissues like the liver, while the SCN remained largely anchored to the light—dark cycle. The implication was unsettling if you liked tidy hierarchies. It meant that "master clock" did not mean "only clock." It meant that a peripheral organ could be re-timed by food even when the brain clock kept counting sunrise and sunset as if nothing had happened.

Why should food be strong enough to do that? Partly because, for the liver, food is not just an event. It is a forecast. A regular breakfast at 8 a.m. is a daily prediction that glucose will soon arrive. A regular dinner at 7 p.m. is a daily prediction that lipid traffic and insulin signaling will soon change. Anticipation is one of the quiet marvels of circadian physiology. We often notice it as hunger that arrives before the meal does, or sleepiness that arrives before bedtime. The liver has its own anticipatory logic, encoded in which enzymes are prepared, which transporters are waiting at the cell surface, and which pathways are set to run hot.

The SCN, entrained by light, tends to impose a broad schedule: wake-promoting signals by day, sleep-promoting signals by night, daily rhythms in core body temperature, and time-of-day patterns in hormones such as cortisol and melatonin. But the SCN does not swallow your dinner for you. The liver "listens" to a different set of cues: the rise in insulin that follows eating, the influx of nutrients through the portal vein from the gut, the changing balance of NADH and NAD^+ inside cells, and the levels of metabolites that indicate whether the organism is in storage mode or in withdrawal mode. Insulin is particularly potent as a refeeding signal; mechanistic work has shown that insulin signaling can modulate components of the liver clock machinery, providing a direct molecular route by which "meal time" can adjust "clock time."

Now picture what happens when light time and food time separate. A shift worker eats a substantial meal at 2 a.m. during a break, under bright workplace lighting, then sleeps during the day in a darkened room. A student eats late while studying, then sleeps in the morning and skips breakfast. A new parent eats whenever the baby settles, which may be in fragments across the 24-hour day. Even without changing *what* is eaten, *when* eating occurs reshapes the informational environment of the liver. The organ is not being stubborn. It is being rational in the way a thermostat is rational: it responds to the input it is designed to treat as meaningful.

Misalignment is often described as if it were merely a scheduling annoyance—as if the body were a calendar app that failed to sync. The more accurate picture is a negotiation between competing time cues. When the SCN says, "night," many tissues interpret that as a command to prioritize repair, to reduce glucose uptake, to alter immune surveillance, to change blood pressure regulation, and to shift energy usage. When the liver receives a meal at that same moment, it is asked to do the biochemistry of daytime: move glucose into storage, turn on pathways for lipid handling, increase bile flow, and deal with the oxidative burden that accompanies nutrient processing. The burden is not only energetic. It is informational. A cell that is told to be in two states at once must either compromise or oscillate poorly.

A concrete example makes the clash vivid. Consider glucose tolerance: how efficiently the body handles a standard amount of carbohydrate. For many people, the same meal will produce a smaller and shorter-lived rise in blood glucose in the morning than in the evening. That is not simply because the morning version is eaten with better intentions. It reflects daily rhythms in insulin sensitivity and in the coordinated response of liver, muscle, pancreas, and adipose tissue. If the liver clock has been shifted by consistent late eating, it may be primed to store and release glucose at hours that no longer match the SCN-driven pattern of sleep and activity. The result can look, from the outside, like a person has suddenly become "bad at carbs" at night. From the inside, it looks like two clocks disagreeing about what time it is.

There is a further twist that can be easy to miss: peripheral clocks do not all reset at the same speed. Damiola's work pointed to the liver as relatively quick to follow feeding time compared with other organs. That detail matters because it suggests a period of internal jet lag even when a new routine is stable. If the liver shifts faster, but kidney or heart shifts more slowly, the body temporarily contains multiple time zones. In everyday life, that can mean that the metabolic machinery is never fully synchronized during the transition: appetite cues, glucose handling, lipid transport, and detoxification capacity may each be peaking at slightly different hours. Many people recognize this as the groggy, subtly unwell feeling of travel across time zones, but without the airplane: a physiology that cannot decide whether it is morning or evening.

The language of "uncoupling" can sound technical, but the experience it points to is intimate. Think about the peculiar hunger that appears late at night even when a person has eaten enough calories. It can feel like a craving with a personality: a different appetite from the one at lunch. Part of that experience is cognitive and cultural, but part is physiological. If meals have trained the liver and related metabolic tissues to expect nutrients at night, the absence of food at that trained time can produce a mismatch signal: hormonal and autonomic patterns that nudge the organism toward seeking food. Conversely, if you habitually stop eating earlier, there is often an adaptation period in which evening appetite flares and then, after days, quiets. That pattern is consistent with a clock being reset, not merely a willpower contest being won.

Anecdotes from the early days of chronobiology show how stubborn biological time can be. When researchers began to place animals and humans in environments without obvious time cues—constant dim light, no clocks, no windows—they found that rhythms persisted. Sleep and wake drifted, body temperature cycled, hormones rose and fell. The body carried its own timekeeping apparatus into the cave. What later feeding experiments revealed was that the cave had more than one clock. You could keep the light-based clock in one state and tug the food-based clock in another. The liver, in this sense, behaves like a timekeeper that can be bribed: not with money, but with nutrients.

None of this makes the SCN irrelevant. The SCN still matters enormously because light is a uniquely reliable signal about the external world. Food can be irregular; light and dark, outside modern cities, are dependable. The SCN coordinates behaviors that powerfully influence feeding: sleep timing, physical activity, and the daily wave of alertness that makes breakfast feel plausible and 3 a.m. feel surreal. But the point of the metabolic shadow clock is that the liver can—and often must—take a local view. From the liver's perspective, a meal is not a philosophical event that should be delayed to honor a theoretical master plan. It is a chemical fact that must be handled quickly, or blood glucose rises, lipids accumulate in the wrong places, and reactive byproducts build up.

A useful way to think about the liver is as an organ that lives at the border between the gut and the body. Blood leaving the intestines passes through the portal vein straight to the liver before it is allowed into general circulation. That anatomical arrangement is a statement of priorities: the liver is a checkpoint, and the first to see what the outside world has delivered. In evolutionary terms, it is easy to imagine why this checkpoint would want its own timekeeping system. If food arrives predictably at a certain time of day, it is advantageous to pre-emptively tune enzyme levels and storage capacity. If food arrives unexpectedly, the liver must respond anyway. A clock that is sensitive to food, rather than only to light, is a clock that can be trained by the most immediate information available.

The modern predicament is that our most immediate information is often self-inflicted. Artificial lighting can stretch the day, and social schedules can stretch eating late into the night. The liver does not know that the late meal is "just a snack" taken in front of a screen, or that the calories were logged in an app and carefully budgeted. It only knows that nutrients have appeared at a time when, by the SCN's reckoning, the body is meant to be fasting. When this happens occasionally, the system can likely absorb it. When it happens nightly, the liver clock is encouraged to shift, and a stable internal disagreement becomes the new normal.

It is tempting to moralize this, to treat meal timing as a personal virtue or vice. A better stance is mechanical sympathy. The liver is trying to solve a scheduling problem with real constraints: it cannot pause glucose processing because the brain thinks it is nighttime. It cannot refuse to package lipids because melatonin is rising. It is an organ that has to do its job when the work arrives. The cost of making it work against the prevailing circadian context is paid not as guilt but as physiology: altered glucose excursions, altered lipid handling, and a body that feels, in subtle ways, less resilient.

One last image helps: think of the liver clock as a lantern carried through a dark street. The SCN is the streetlight grid, turning on and off with the city. When the lantern is carried in step with the streetlights, it adds illumination where you need it. When it moves to a different rhythm, it does not go out—it simply casts light in the wrong places, at the wrong times, and your shadows fall in directions you do not expect.

4.2 Host and Guest at the Table

A late-night meal is often described as a private decision: a bowl of cereal eaten in silence, a slice of pizza folded over the sink, a sweet snack unwrapped with the guilty speed of someone hoping not to be noticed. The body does notice. So do the trillions of microbial cells in the gut, which respond to food with the enthusiasm of a crowded dining room hearing the kitchen bell. Their appetite is not a metaphor. It is chemistry. And like any good dinner party, timing determines whether the evening feels convivial or chaotic.

The gut microbiome is the shifting community of bacteria, archaea, fungi, and viruses that live along the digestive tract, especially in the colon. "Community" is not sentimental language here. These organisms interact with each other, compete for resources, and produce metabolites that influence their neighbors. They also interact with the host, and that relationship is not only about digestion. The gut microbes help break down otherwise indigestible fibers, train the immune system, produce vitamins, and generate a fog of small molecules that enter the bloodstream and reach the liver, fat tissue, and brain. The host, for its part, supplies warm housing, steady hydration, and an endlessly renewable buffet of mucus, dietary residues, and shed intestinal cells. Host and guest are so interdependent that it is hard to say where one ends and the other begins.

Yet there is a boundary, and the boundary keeps time.

The inner surface of the intestine is not bare flesh. It is a living barrier, made of a single layer of epithelial cells stitched together by protein complexes called tight junctions. Over that cellular layer lies mucus, a gel rich in sugars (mucins) secreted by specialized cells. Within that mucus lives a huge fraction of the microbiome, held at a controlled distance from the epithelium. This arrangement lets the host harvest microbial benefits while reducing the risk of microbes crossing into tissue where they would be treated as invaders. The barrier is therefore a kind of *bouncer* at the party door: friendly to those who behave, quick to eject those who get too close.

Barrier function is not constant over the day. It changes with sleep and wake, with feeding and fasting, and with the daily rhythm of hormones and immune signals. The details differ by species and by experimental conditions, but the organizing idea is consistent: the gut is not equally open for business at all hours. During the active phase (daytime for humans), the intestine is primed for the predictable work of eating: absorption of nutrients, regulated sampling of microbial products to educate the immune system, and an infrastructure that tolerates the bustle. During the rest phase, the intestine shifts emphasis toward maintenance: renewing epithelial cells, reorganizing mucus, and repairing microscopic wear. Even if a person does not notice these changes directly, the microbiome does, because its world is built on what seeps through.

The simplest signal the microbes receive is also the one most likely to be disrupted: the timing of food. When meals arrive at roughly consistent hours, nutrient availability in the gut has a daily rhythm. When meals are scattered across the day and night, nutrient availability becomes more constant. Many gut bacteria can proliferate rapidly when their preferred substrates appear, and many can switch metabolic modes when substrates disappear. If you could shrink yourself to the size of a bacterium and watch a single day in the colon, you would not see a stable landscape. You would see tides.

Those tides are not only about bulk calories. They are about *which* molecules arrive. A meal rich in simple sugars changes the environment differently from a meal rich in fiber. Protein residues feed a different set of microbes than complex carbohydrates do. Fat changes bile secretion, and bile acids are both detergents and signaling molecules that can inhibit some microbes while favoring others. The microbiome is therefore not merely "fed" or "fasted"; it is continually being steered toward different ecological states. The surprising part is that this steering can become rhythmic, like a commuter train schedule, with predictable surges in certain taxa and predictable bursts of certain metabolic pathways.

One reason this matters for circadian biology is that microbial metabolites are time-stamped messages. Short-chain fatty acids (SCFAs) such as acetate, propionate, and butyrate are produced when microbes ferment fibers. These molecules can nourish colon cells, influence gut motility, and affect glucose and lipid metabolism once absorbed. Other microbial products shape the immune system and alter signaling in the nervous system. If microbial production of these compounds oscillates across the day, then the host receives an oscillating chemical input that can help synchronize peripheral clocks, including in the liver. If microbial rhythms are flattened or shifted, the host receives chemical signals at unfamiliar times. The party goes on, but the music is wrong for the hour.

The circadian character of the microbiome became difficult to ignore after experiments showed daily oscillations in microbial composition and function. Work by Thaiss and colleagues (2014) was particularly influential, not because it romanticized the microbiome but because it treated it as a dynamic system embedded in the host's circadian network. They reported that microbial populations and their gene activity oscillate over the day, and that disrupting the host circadian system or imposing jet lag disrupts these microbial oscillations.

More strikingly, the metabolic consequences of that disrupted microbial community could be transferred: microbes taken from jet-lagged hosts and transplanted into germ-free mice promoted weight gain and glucose intolerance. The message was uncomfortable in the best scientific way. It suggested that some of what we experience as "metabolic consequences of circadian disruption" may be mediated through an altered microbial ecosystem, not solely through host hormones and host tissues.

Transplant experiments have a long, oddly theatrical history in biology. In the early twentieth century, scientists used tissue transplants and parabiosis (surgically joining animals to share circulation) to test whether physiological states were local or systemic. Those experiments can be unsettling to modern eyes, but they produced a key conceptual tool: if you can transfer a trait by moving a biological component from one animal to another, you have evidence for causality that correlation alone cannot provide. In microbiome research, germ-free mice play this role. They are not a perfect model of human life, but they are a powerful way to test whether microbes can carry metabolic tendencies. The jet-lag microbiome transplant fits into this tradition: it is an argument, made with living evidence, that timing can propagate through biology by way of other organisms.

Timing, here, means not only the SCN's light-based schedule but also the feeding rhythm that the liver clock cares about so deeply. When meals consistently occur during the active phase, feeding rhythms become a strong organizer of microbial rhythms. When meals occur during the rest phase, the microbes are asked to do their peak work at a time when the host barrier and immune system may be in a different mode. The host can adapt to occasional disturbances, but persistent late eating changes the ecological game.

A concrete scenario makes this less abstract. Consider someone who eats dinner at 9:30 p.m., snacks at 11:30 p.m., and then sleeps around 12:30 a.m. The intestine receives nutrients late. Bile acids are secreted late. Gut motility patterns that usually slow during sleep now overlap with active digestion. The microbiome, which had been drifting toward a fasting state earlier in the evening, is abruptly given new substrate. Fermentation and microbial growth rebound. Meanwhile, the barrier may be shifting toward its nighttime maintenance program, with different patterns of mucus secretion and tight junction regulation. The result is not necessarily a dramatic breach, but a subtle increase in the likelihood that microbial products cross the boundary when the immune system would rather not be disturbed. This is one plausible route by which late-night eating can feel inflammatory over time: not because the food is "toxic" in itself, but because it arrives during a physiological window that is less tolerant.

People sometimes hear the phrase "leaky gut" and imagine a torn hose: an obvious rupture that spills bacteria into the bloodstream. Real biology is subtler and, in some ways, more worrying. Intestinal permeability exists on a spectrum. Tight junctions are not welded seams; they are adjustable. The immune system does not treat every microbial molecule as a crisis; it constantly samples and interprets. Many microbial products can cross in small amounts without causing harm. Problems arise when the timing, quantity, or context of that crossing changes—when the boundary is asked to do too many things at once, too often, at the wrong hour.

The gut also sits under the influence of the nervous system. The enteric nervous system, sometimes called the "second brain," coordinates local reflexes and communicates with the central nervous system via the vagus nerve and hormonal signals. Stress can change gut motility, secretion, and permeability; sleep loss can do the same. This matters because circadian misalignment rarely arrives alone. Night work, jet lag, or chronic late-night screen time often come with shorter sleep, more stress, and different food choices. Untangling these factors is hard in real life, and careful studies are needed to isolate meal timing from everything else. Still, the biological architecture makes one point hard to escape: the gut is a meeting place where behavioral time meets immune time and microbial time.

A useful metaphor is the distinction between a dinner party and a diner that never closes. In a dinner party, there is a clear arc: guests arrive, food is served, conversation swells, and then the evening winds down. The host can clean up, reset the room, and prepare for the next event. In a diner that never closes, the kitchen is always half-open, and the staff is always half-cleaning while half-serving. Many modern eating patterns resemble the diner: breakfast that bleeds into a mid-morning snack, lunch followed by an afternoon bite, dinner followed by dessert followed by "just something small" before bed. Even if total calories are not excessive, the gut and its microbes lose the clean separation between feeding and fasting that helps keep rhythms sharp. A rhythmic microbiome becomes a smeared microbiome, and a barrier that expects quiet is asked to keep performing.

One place where this becomes socially visible is in the lives of shift workers. Hospitals, warehouses, emergency services, and transport networks depend on people being awake and eating at night. These workers often report gastrointestinal symptoms: heartburn, irregular bowel movements, abdominal discomfort. Some of that is due to sleep disruption and stress, but meal timing is a particularly direct lever. Eating a substantial meal at 3 a.m. is not merely a behavioral oddity. It is a signal that propagates through bile secretion, microbial fermentation, and barrier dynamics. Laboratory simulations of night work have shown that restricting eating to the daytime can reduce internal circadian misalignment and improve glucose responses compared with eating at night. A reasonable inference, consistent with what is known about microbial rhythms, is that daytime eating during night work may also help preserve more normal oscillations in the gut ecosystem, even if sleep timing is shifted.

The symbiosis between host and microbes also has a diplomatic dimension. The immune system has to decide when to tolerate and when to attack. The gut is full of bacterial molecules that, in other contexts, would scream "infection." Lipopolysaccharide (LPS), a component of the outer membrane of many bacteria, can provoke strong inflammation if it enters the bloodstream in sufficient amounts. The body has mechanisms to neutralize and clear such molecules, but these mechanisms are themselves rhythmic. If permeability increases at a time when clearance is not optimized, then the same microbial exposure can be more provocative. Here again, timing is not a footnote. It is part of the risk calculation.

This is not an argument for living in fear of one's microbes. It is an invitation to recognize them as partners whose behavior is shaped by the same cues that shape ours. When people talk about "listening to your gut," they often mean intuition or mood. There is also a literal sense in which the gut is listening: to the regularity of meals, to the daily pause that comes with fasting, to the hormonal rhythm of sleep and wake. When that listening is rewarded with consistent cues, the microbial community tends to settle into predictable oscillations, and the barrier can coordinate its maintenance around periods of relative quiet. When the cues become erratic, the microbiome becomes more like a crowd that never quite leaves the house.

A small but telling everyday example is the difference between waking up hungry and waking up nauseated. Many people who eat very late report diminished morning appetite, sometimes even queasiness. This can be dismissed as simple indigestion, but it also makes sense as a mismatch between feeding and the gut's morning state. Overnight, the intestine typically transitions through phases of motility and secretion that prepare for the first meal of the day. If the previous night's meal pushed digestion deep into sleep, the morning may begin with a gut still managing remnants, with altered bile patterns and microbial activity. Hunger is not only about calories needed; it is about a body confident that it is ready to process food.

The microbiome's rhythms also remind us that circadian networks are not purely human. They are multi-species. The host sets the table: meal timing, mucus secretion, immune tone, intestinal motility. The guests respond: growth, fermentation, metabolite production. Those metabolites feed back into the host: influencing liver metabolism, appetite, and immune signaling. There is no single master in this arrangement. There is coordination, and when coordination breaks down, the costs can appear as foggy symptoms long before formal disease.

A satisfying meal at the right time often brings a particular kind of calm: not just fullness, but a sense that the body has settled. Part of that calm may be social and psychological. Part of it is also the quiet work of a well-timed truce at the gut boundary, where the host can absorb what it needs and the microbes can feast without pressing too hard against the door.

4.3 The Fasting Reset

Time-restricted eating has the clean appeal of a household rule. Eat within a window. Close the kitchen. Stop negotiating with yourself at 10:47 p.m. over whether yogurt counts as dinner or dessert. For many people the attraction is not only metabolic; it is moral and mental, a relief from constant decisions. That simplicity is also why the idea is repeatedly sold as a miracle: a way to lose weight without trying, to "heal inflammation," to reverse aging, to outsmart biology with a schedule.

Biology is not impressed by slogans, but it does pay close attention to the difference between being fed and being fasted. The most useful way to approach time-restricted eating (TRE) is therefore not as a diet but as a manipulation of *state*. It changes the daily alternation between anabolic physiology (the building-and-storing mode that follows meals) and catabolic physiology (the breaking-down-and-repair mode that emerges as fasting deepens). The question is not whether TRE is magical. The question is what changes when the body is allowed a reliable fasting interval, and what changes when that interval is eroded by grazing.

Start with a scene that is so ordinary it rarely feels like an experiment: a person who eats breakfast at 8:00 a.m., lunch at 1:00 p.m., dinner at 7:00 p.m., and nothing after. Compare that with a person who eats the same total calories but spreads them from 8:00 a.m. until midnight, with small snacks and sweet drinks. Both people may claim to "eat all day," but physiologically they are living in different countries. The first person has a long, quiet overnight stretch in which insulin falls, the liver shifts from storing glucose to releasing it, and cellular housekeeping processes become easier to run without interruption. The second person keeps sending small signals of feeding into the system, repeatedly nudging insulin upward and repeatedly restarting digestion and nutrient processing. The body can still cope, but it never quite gets the message that the shop is closed.

This is where hype often begins: with the true statement that fasting intervals can activate pathways associated with maintenance, and then the leap that *any* eating window will automatically improve health for *everyone*. The truth is more specific, and more interesting. Timing interacts with the circadian network described earlier: the light-entrained brain clock, the food-entrained liver clock, and the microbial and barrier rhythms at the gut boundary. TRE can be beneficial not only because it reduces calories (sometimes it does, sometimes it does not), but because it can reduce internal disagreement about time by making the daily pattern of feeding and fasting more predictable.

A common misunderstanding is that TRE is merely "skipping breakfast" or "not eating after 8." Those behaviors can fall under TRE, but they are not equivalent. The clock matters. Eating the same eight hours earlier in the day is physiologically different from eating eight hours later, even if the fasting duration is identical. One reason is glucose handling: many people have better glucose tolerance in the morning than in the evening. Another reason is that the hormones and autonomic signals that accompany alertness and sleep are not neutral backdrops; they shape how tissues respond to meals. In the language of circadian biology, a meal is not only calories. It is a *zeitgeber*—a time cue—for metabolic organs.

The strongest early mechanistic push for TRE came from animal work that separated timing from calories with unusual clarity. In 2012, Satchidananda Panda's group (Hatori and colleagues) showed that mice fed a high-fat diet ad libitum tended to spread their intake across the day and night, blunting normal feeding rhythms. When the same calories were constrained to a consistent daily window, many metabolic harms were reduced. The result was not that clocks became perfect, but that rhythms sharpened: oscillations in metabolic pathways and clock-controlled genes looked more like those of healthier animals. In mice, with their exquisitely controlled environments and obedient diets, this was a compelling demonstration that *when* can matter even when *how much* does not change.

Humans are not mice, and it would be irresponsible to pretend otherwise. Human studies of TRE show variability: some protocols produce clear improvements, others produce modest or inconsistent effects, and adherence often determines outcomes as much as biology does. Still, a handful of controlled trials have been illuminating because they minimized confounders. A particularly discussed example is early time-restricted feeding (eTRF), where eating is shifted earlier in the day and dinner becomes small or absent. Elizabeth Sutton and colleagues (2018) studied men with prediabetes and found that eTRF improved insulin sensitivity and other cardiometabolic markers even without weight loss. That detail matters because it pushes against the simplistic claim that TRE "works" only by reducing calories. In a matched-intake setting, timing itself appeared to change physiology.

Follow-up work by Jamshed and colleagues (2019) added nuance: under controlled conditions, eTRF did not boost total 24-hour energy expenditure, but it did alter appetite and substrate use, increasing fat oxidation and making hunger more stable. Here again, the lesson is neither miracle nor placebo. The body did not burn extra calories out of thin air. It changed *what it was burning* and *how it felt*, which can matter enormously over weeks and months in the messy reality of free-living humans.

What might explain these effects? Part of the answer lies in the liver's sensitivity to insulin as a clock-setting signal. When meals arrive, insulin rises and activates signaling pathways (including Akt) that communicate "fed state" to cells. Mechanistic studies have shown that insulin signaling can influence components of the liver clock machinery, including BMAL1, providing a molecular link between meal timing and hepatic timekeeping. If a person eats late every night, insulin peaks late every night, and the liver clock is repeatedly encouraged to shift. If a person consolidates eating earlier, insulin peaks earlier and the liver clock can align more readily with the SCN-driven sleep phase. TRE, especially in an earlier window, can therefore be understood as a way of reducing mixed messages: fewer insulin spikes during the biological night, fewer metabolic instructions issued when the brain is issuing sleep instructions.

Insulin is not the only actor. The fed-to-fasted transition involves a cascade of changes. Glycogen in the liver is gradually depleted; the body becomes more reliant on fatty acids; ketone bodies begin to rise as fasting deepens. One of these ketones, β-hydroxybutyrate (βOHB), has received attention not only as a fuel but also as a signaling molecule that can influence gene regulation. In other words, fasting does not merely change what is burned; it changes what messages circulate.

Ketones also highlight a subtlety that hype tends to erase: different dietary patterns can produce different rhythms of the same molecule. Caloric restriction, time restriction, and ketogenic diets can all increase βOHB, but they can do so with different amplitudes and different timing. Some work suggests that caloric restriction produces particularly high-amplitude daily ketone rhythms, while time-restricted feeding can also induce rhythmic ketones but sometimes with smaller swings, depending on the protocol. Meanwhile, ketogenic diets can markedly reshape liver metabolic outputs and the timing of liver programs, in part because the diet continuously biases the body toward fat-derived fuels. The implication is not that one pattern is universally superior. It is that the liver's clock and metabolic machinery are plastic, and the *pattern* of signals matters as much as their average level.

This helps clarify what "reset" can and cannot mean. The body is not a laptop that can be rebooted into factory settings after a 16-hour fast. The clockwork is always running, and its phase is shaped by repeated cues over days. But there is a real sense in which a reliable fasting interval can restore contrast. In circadian terms, contrast is amplitude: the difference between peak and trough. A physiology that is always half-fed is a physiology with reduced amplitude. Signals that should rise sharply after a meal and fall decisively overnight become rounded hills. Signals that should be quiet at night remain noisy. A consistent fasting period can sharpen those contrasts, giving tissues a clearer sense of "now we store" and "now we repair."

Repair is another word that invites exaggeration. Many cellular maintenance processes—autophagy is a famous example, a form of intracellular recycling—tend to be more active when nutrients are scarce. But the relationship is not a simple on/off switch controlled by the clock app on your phone. Autophagy occurs in fed states too; it is regulated by many factors including nutrient sensors such as mTOR and AMPK, stress signals, and tissue-specific demands. TRE may encourage a daily interval in which catabolic and recycling programs can run with fewer interruptions, but the magnitude and health relevance of that shift likely depend on baseline metabolic health, sleep, exercise, and the actual length and consistency of fasting achieved.

This is where diet-industry claims often become both seductive and scientifically sloppy. "Fasting turns on autophagy" is true in a broad, qualitative sense, but it is not a promise that a particular eating window will create cellular cleansing powerful enough to override chronic sleep loss, sedentary behavior, or a highly processed diet. TRE can be a lever. It is not a substitute for every other lever.

One reason TRE became culturally resonant is that it echoes older patterns of eating that were not branded as interventions. Before electric lighting and 24-hour food access, many communities had a naturally longer overnight fast simply because kitchens were closed and food preparation took work. Religious fasting traditions also created intermittent periods of food restriction, though for reasons that were spiritual rather than metabolic. Those traditions offer a historical anecdote worth holding onto, not as proof of health benefits, but as evidence that humans have long experimented with time as a dimension of eating. The modern novelty is not the existence of fasting. It is the near-disappearance of fasting in everyday life, replaced by constant availability and the normalization of late-night calories.

The gut microbiome adds another layer to the "reset" idea. Many gut microbes respond quickly to nutrient availability, and the daily pause of fasting changes which substrates are available in the colon. A reliable fasting interval can promote more consistent microbial oscillations by creating a predictable daily rhythm in what reaches the gut and when. That, in turn, may influence the production of microbial metabolites that affect host metabolism. This is still an emerging area in humans, and it would be premature to claim that TRE reliably "fixes" the microbiome. But it is reasonable to view feeding rhythms as a system-level organizer that can either support or disrupt host–microbe coordination.

If all this makes TRE sound complicated, it is because the body is complicated. The practical take-home is surprisingly plain: the metabolic system seems to benefit from having a clear stretch of time each day when it is not being asked to process incoming calories. Whether that stretch is 12 hours, 14 hours, or occasionally longer is likely less important than whether it is consistent and aligned with sleep for that individual. For a day-shift worker who sleeps at night, an earlier eating window often aligns better with the body's natural glucose-handling rhythm than a late one. For a night-shift worker, the situation is harder, because light exposure, sleep timing, and meal timing are all under pressure from the demands of work. In that context, even partial alignment—such as concentrating meals during the daytime rather than during the biological night—may reduce internal conflict.

The most honest view of TRE avoids two traps. The first trap is cynicism: dismissing meal timing as a fad because some claims are exaggerated. The second trap is evangelism: treating meal timing as a universal cure. Between those extremes lies the physiology: clocks in the liver and other tissues that pay attention to food, microbial ecosystems that surge and ebb with nutrient flow, and a barrier that appreciates quiet hours. A consistent fasting interval does not make the body perfect. It makes the signals cleaner, the transitions more decisive, and the conversation between host and metabolism a little less like people talking over each other at the end of a long night.

Chapter 5

The Gate of Sleep

Sleep is often misunderstood as a simple lack of wakefulness, but it is actually a precisely timed collision between accumulating metabolic debris and an internal alerting rhythm. We will dismantle the mechanics of drowsiness to reveal how two opposing biological forces negotiate the boundaries of consciousness, determining not just if we sleep, but exactly when the body permits it.

5.1 The Collision of Two Forces

By late afternoon, many people can still answer emails, drive a car, and carry on a conversation without much trouble, yet feel a kind of invisible sandbagging in the background: attention frays, patience thins, and the simplest tasks start to feel heavier than they did at noon. If sleepiness were nothing more than a steadily filling tank—more hours awake, more urge to sleep—then the story would be simple. You would get progressively, predictably sleepier, and the evening would be the worst of it.

But real evenings often refuse to cooperate with that tidy picture. There is the familiar *second wind*: the moment when a person who was dragging at 6 a.m. suddenly feels oddly capable at 9 p.m. There is the equally familiar frustration of trying to fall asleep early before a flight, a shift, or an exam, only to lie there with an alert mind that seems to have misplaced the very concept of sleep. And there is the paradox that a person can be profoundly sleep deprived and still find themselves wide awake at precisely the wrong times.

A widely used way to make sense of these contradictions is the Borbély two-process model of sleep regulation, developed in the early 1980s and refined in collaboration with others. It is not a claim that the brain has only two parts relevant to sleep, or that sleep is governed by two molecules and nothing else. It is a deliberately simple formal model: two forces with different shapes in time, whose interaction predicts when sleep is likely to happen and how deep it tends to be once it starts. Its power lies in how much everyday experience it explains with so little machinery.

One force is a homeostatic drive—a word borrowed from physiology, meaning a stabilizing tendency that pushes a system back toward balance. Here the "balance" is not blood sugar or body temperature, but the balance between time spent awake and time spent asleep. The longer you are awake, the stronger this drive becomes. In the model it is called *Process S*, with the "S" often read as "sleep" or "slow waves," because one of its best physiological correlates is the amount of slow-wave activity in the sleeping brain.

The other force is circadian—from *circa diem*, "about a day"—and it behaves less like a filling tank and more like a tide. Across the 24-hour cycle, it provides a wake-promoting push at some times and relaxes at others. In the two-process model it is called *Process C*, and it is often drawn as a smooth oscillation, a daily rise and fall that is anchored, in ordinary life, to environmental time cues such as light.

On its face, this might sound like a mere naming of what everyone knows: you get sleepier with time awake, and nights are different from days. The insight is that these two influences are not redundant. They can oppose each other strongly, and that opposition is what creates the peculiar topography of human sleepiness: its peaks, its lulls, its awkward timing, and its gates.

To make the logic concrete, imagine a typical day for a person on a regular schedule. They wake around 7 a.m. and remain awake until 11 p.m. If Process S were the whole story, their sleep pressure would climb steadily through the day and reach its maximum at 11 p.m. That part is roughly true. Yet subjective sleepiness often does not climb in a straight line. Many people feel a dip after lunch. Then, despite accumulating fatigue, they rally into the evening, sometimes feeling most alert precisely when they think they *should* be preparing for bed. This evening rally is not a triumph of willpower; it is a feature of the timing system.

Process C, in this framework, acts like an internal wake-maintenance signal that is strongest in the late biological day. It counterbalances rising Process S and makes it possible to stay awake for a consolidated daytime episode rather than sliding into random naps whenever the homeostatic drive becomes noticeable. In other words, the circadian system does not merely *permit* sleep at night; it *prevents* sleep at certain times of day. That prevention can be so effective that sleep becomes remarkably difficult even when you are tired.

Peretz Lavie famously described an evening "forbidden zone" for sleep in the 1980s, based on protocols that sampled sleep propensity across the day using repeated opportunities to fall asleep. The label is vivid and slightly mischievous, but it captures a common experience: in the early evening, sleep can feel strangely inaccessible. It is not that Process S has stopped rising. It is that Process C is pushing hard in the other direction. The body is, in effect, keeping the gate shut.

Then, as the night approaches, Process C changes character. Its wake-promoting influence diminishes, and the same level of homeostatic pressure that was held at bay in the evening can suddenly become decisive. The gate opens. Sleep onset, which felt impossible at 9 p.m., can become effortless at 11 p.m. or midnight. In many people, that transition can feel abrupt, almost binary: "I could not sleep" becomes "I can barely keep my eyes open." The two-process model gives that abruptness a simple geometric interpretation: it is the moment when the rising curve of Process S finally overtakes, or is no longer opposed by, the circadian alerting signal.

The homeostatic process is often linked, in popular descriptions, to the molecule adenosine. Adenosine is a breakdown product of cellular energy use, and it accumulates in certain brain regions during prolonged wakefulness. It is not a poetic metaphor to say that wakefulness leaves chemical traces; the brain is metabolically expensive, and sustained activity shifts the balance of neuromodulators. Caffeine, the world's most widely used psychoactive substance, acts in large part by blocking adenosine receptors. That is why coffee can temporarily push back sleepiness without "adding" energy in any literal sense: it interferes with the brain's ability to *feel* the homeostatic load.

Even so, Process S should not be thought of as a single molecule. Homeostatic pressure is an abstraction, a summary of many local and global factors: synaptic activity, metabolic state, neuromodulatory tone, and the networks that regulate arousal. The reason slow-wave activity (those large, slow oscillations seen in deep non-REM sleep) is often used as a marker is that it behaves in a distinctly homeostatic way: it is higher after long wakefulness and declines across the course of a night's sleep. That decline is the signature of dissipation, the mathematical mirror image of the buildup across the waking day.

Process C, meanwhile, is the timed influence of the circadian pacemaker and its downstream outputs. Importantly, it is not merely a sleep timer that says "now is night." It is better imagined as a *threshold setter* and a *state stabilizer*. In the classic extension by Daan,

Beersma, and Borbély in the 1980s, sleep onset and offset can be pictured as occurring when Process S crosses boundaries that move with circadian phase. Those boundaries—upper and lower thresholds—shift across the day in such a way that sleep is encouraged at night and discouraged in the late afternoon and evening. The effect is to consolidate sleep into a single, efficient block and to keep wakefulness coherent during the day. Without such gating, the homeostatic drive might produce a more fragmented pattern: dozing whenever pressure becomes too high, waking whenever it eases, with less predictable timing.

A useful everyday demonstration comes from an experiment many people unintentionally perform on themselves: the weekend sleep-in. Suppose a person keeps a demanding weekday schedule, waking at 6:30 a.m. and falling asleep near 11:30 p.m. By Friday they have accrued some sleep debt: Process S has been kept chronically high, never fully paid down. On Saturday they sleep until 10 a.m. Process S drops more than usual because sleep lasted longer, and they feel restored. The temptation is to assume that bedtime will naturally shift earlier because they are "caught up." But Saturday night often moves later, not earlier. The reason is that Process C is still running on a roughly 24-hour rhythm and is sensitive to light and timing. Sleeping late changes morning light exposure and pushes circadian phase later in many people. Now bedtime arrives, but the circadian alerting signal is still buoyant. The person feels alert at midnight, falls asleep at 1:30 a.m., sleeps late again, and by Monday morning feels jet-lagged without crossing a time zone. The model's language for this is simple: Process S may be lower because of recovery sleep, but Process C has shifted, moving the sleep gate later.

The same interplay explains why a nap can be both a relief and a trap. A short nap in the mid-afternoon often helps because it chips away at Process S at a time when Process C is not maximally wake-promoting. Many people have a natural dip in alertness after lunch, and in Lavie's measurements there is often a secondary peak of sleep propensity in the afternoon. Yet a long nap late in the day can reduce Process S enough that, when the circadian gate finally opens at night, there is simply not enough pressure left to fall asleep quickly. The person is not "wired" in some moral sense; they are under-pressured at the wrong circadian phase.

Shift work makes the collision between these forces painfully visible. Consider a nurse who must work from 11 p.m. to 7 a.m. On the first night shift, Process S rises through the night as it always does during wakefulness, but Process C is doing the opposite of what the worker needs. In the early night, circadian wake promotion may still be relatively strong, making it possible to begin the shift. But as the biological night deepens, Process C withdraws its wake-promoting support precisely when the homeostatic load is now high. The combined effect is a trough in alertness at the worst possible time, often near the circadian low point in core body temperature. Then, in the morning, when the shift ends, Process S is maximal and the worker wants sleep—but Process C is beginning to rise, providing a wake-promoting push that fights daytime sleep. Blackout curtains, eye masks, and rigid routines are attempts to manipulate the circadian input (especially light) and to protect the fragile alignment between the two processes. Many shift workers live in a permanent compromise: enough circadian adaptation to function, but not enough to fully move the sleep gate to the day, especially when days off demand a return to normal social time.

Laboratory studies designed to separate these processes have underscored the point. In forced desynchrony protocols, participants live on an artificial "day" length that the circadian system cannot follow, such as 28 hours. Sleep opportunities occur at every circadian phase, sometimes when the internal clock is urging wakefulness, sometimes when it is urging sleep. Under those conditions, sleep does not simply mirror time awake. It shows clear circadian modulation: the ease of falling asleep, the continuity of sleep, and the architecture of sleep stages vary with circadian phase even when homeostatic pressure is matched. REM sleep, in particular, shows a robust circadian rhythm, clustering at specific circadian times. Meanwhile, markers tied to Process S, such as slow-wave activity, track prior wakefulness strongly and show only modest circadian influence. The two influences are separable in principle, even if in daily life they are braided together.

Behind the elegant curves sits an important psychological reality: people are poor judges of the underlying forces. We often interpret evening alertness as evidence that we are "not tired" or that we can safely borrow more time from sleep. Yet the model predicts that evening can be misleading. During the forbidden zone, a strong circadian alerting signal can mask high homeostatic pressure, creating the impression that sleep is unnecessary. When the circadian signal drops later, the mask falls away. The crash is not sudden laziness; it is a change in the balance of two drives.

That masking also explains why insomnia can be so perplexing. Someone may climb into bed at 9:30 p.m. because they are exhausted, only to lie awake with an active mind. If their circadian system is delayed—by evening light, irregular schedules, or an innate tendency toward later timing—then their sleep gate may simply not be open yet. Homeostatic pressure is present, but it is being actively countered. Telling such a person to "relax" is like advising someone to lower the ocean with positive thinking. Conversely, another person may fall asleep rapidly at 9:30 p.m. not because they have superior sleep hygiene but because their circadian phase makes the gate earlier.

The model also helps clarify a common misconception: that sleep onset is purely a matter of accumulating enough fatigue. If that were true, a person who stayed awake all night would necessarily fall asleep instantly at 8 a.m. the next morning. In fact, many people find that after the worst of the night they feel oddly alert in the morning. Part of that is psychological (the world is waking up; the brain has novelty to feed on), but part is biological: Process C is rising, lending wakefulness a scaffold. The homeostatic drive is high, and the person may experience microsleeps or lapses, yet there can be a brittle, unnatural form of alertness. It is not healthy; it is the circadian system doing its job at the wrong time for the person's needs.

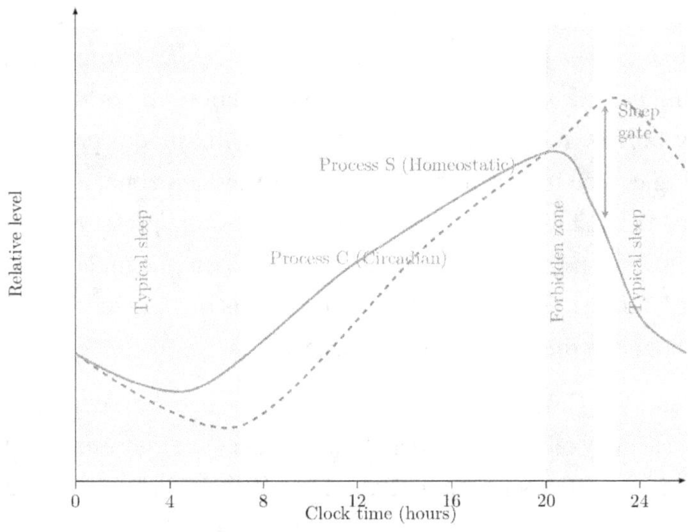

The picture is not meant to be a precise measurement for any one person. It is a conceptual map, a way to see why willpower alone cannot set sleep on command. During much of the day, Process C holds wakefulness up, even as Process S climbs. In the evening forbidden zone, that support is often near its strongest. The distance between the two curves becomes large: high homeostatic pressure, high circadian wake drive. People can feel strangely alert and yet oddly brittle, as if attention could shatter if struck. Then the circadian signal drops. The same accumulated pressure that was successfully resisted becomes effective. The gate opens.

There is a further implication, quietly radical for how many modern lives are organized. Because Process C is sensitive to light and timing, it can be pushed later by bright evenings and pulled earlier by bright mornings. Artificial light does not merely extend the day in a social sense; it can shift the biology that decides when the gate opens. A person who spends the evening under strong indoor lighting, scrolling on a bright screen, may feel capable late into the night not only because they are entertained or stressed, but because the circadian system is receiving a "daytime" message. Meanwhile, the homeostatic pressure continues to climb, unseen. When the alarm rings, Process C may not yet be providing its morning lift, and Process S may not have been adequately paid down. The result is the specific misery of an early alarm after a late night: a sense that the body is being asked to be awake at a time it has not authorized.

The two-process model has survived not because it captures every nuance, but because it captures a central truth about sleep: timing is not the same thing as tiredness. Sleep happens when physiology grants access. The gate is the moment those forces stop canceling each other and begin to cooperate, and when it finally opens, the brain does what it evolved to do with remarkable speed and decisiveness: it lets go.

5.2 Hardwired Horizons

At breakfast, people tell stories about themselves that sound like moral confessions. "I'm useless before coffee." "I wake up at 5 a.m. without trying." "I come alive at 10 p.m." These claims often get filed under personality: disciplined or chaotic, ambitious or lazy, a "morning person" with their life together or a "night owl" who needs to grow up. Yet the most stubborn feature of these stories is how little they respond to resolve. Someone can train for a marathon, quit smoking, learn a language, and still find that a 6 a.m. alarm feels like an assault on the nervous system. Another can swear they will sleep in on vacation and still snap awake at dawn, eyes open, brain already running its first errands.

Chronotype is the neutral name for this tendency: an individual's characteristic timing of sleep and wakefulness, along with the daily rhythm of alertness that makes certain hours feel naturally bright or dim. It is not a label for habits (though habits can mask or exaggerate it). It is a phenotype, a measurable variation in biological timing that shows up in questionnaires, in sleep logs, in melatonin timing under dim light, and in how quickly the internal clock shifts when the external world demands it. The striking thing is not that people differ—human beings differ in everything—but that the difference has the feel of a horizon: an edge you can approach, negotiate with, sometimes lean against, but rarely cross without cost.

The two-process picture of sleep timing makes this immediately legible. Sleep begins most easily when the homeostatic pressure has built up and the circadian system has stopped propping wakefulness. In a person whose circadian phase is earlier, the circadian "permission" for sleep arrives sooner and the morning lift arrives sooner. In a person whose circadian phase is later, the evening wake-maintenance signal stretches deeper into the night, and the biological morning begins later too. The homeostatic drive still rises with time awake for everyone, but the *gate* moves. A chronotype is, in a practical sense, the placement of that gate relative to the social clock.

This is why advice that treats sleep as purely homeostatic so often fails. "Just wake up earlier and you'll be tired earlier." Sometimes it works, briefly. But it often produces a miserable hybrid: an earlier alarm that raises sleep pressure and makes the day foggier, paired with a circadian system that continues to insist on a later night. The person becomes both tired and wired, and then blamed for being neither productive nor well-rested. Habit matters, but it operates on top of a timing system that has its own inertia.

A historical episode makes the point with unusual clarity. In the mid-twentieth century, sleep researchers began to study humans in environments stripped of time cues: no clocks, no daylight, no social schedule. One of the best-known series of studies was conducted by Jürgen Aschoff and colleagues, and later by others who used underground bunkers and caves as natural laboratories. Participants were allowed to choose when to sleep and wake. Many drifted to a daily cycle slightly longer than 24 hours, a "free-running" rhythm that revealed the internal clock's natural period. But what startled observers was how much people differed. Some stabilized near 24.2 hours, others longer; some preferred long bouts of wakefulness followed by long sleep; others adopted different patterns. Even in a world without meetings, school bells, or sunrise, people did not converge on a single "human schedule." They expressed their own timing.

That finding has been refined and disciplined by modern chronobiology, but its emotional implication remains: when the external scaffolding is removed, the internal scaffolding does not vanish. It becomes visible.

Chronotype is partly shaped by age. Adolescence tends to push the circadian system later, and early school start times collide with that biology with predictable results: shortened sleep, morning sleepiness, and a kind of chronic misalignment that can look like attitude. Older adulthood often shifts earlier again. These are population trends, not rules; there are teenage larks and elderly owls. But the age pattern is strong enough that it should have changed how society interprets teenagers' mornings long ago.

Chronotype is also partly shaped by latitude, season, and exposure to natural light. Bright morning light tends to pull the circadian clock earlier; bright evening light tends to push it later. Modern lighting tilts heavily toward the latter. Many people spend their days indoors in comparatively dim light and then encounter bright, blue-enriched light in the evening from screens and overhead LEDs. The internal clock, which evolved to treat light as the most reliable daily cue, does what it is designed to do: it updates its expectations. If a person already has a tendency toward a later schedule, this environment becomes an amplifier.

But the most contentious question—and the one people care about at the dinner table—is whether chronotype is "real" in the sense of being biologically anchored, or whether it is simply the sum of choices. Genetic studies and rare clinical syndromes have made it hard to dismiss chronotype as mere preference.

A dramatic example is familial advanced sleep phase, in which multiple members of a family become sleepy very early in the evening and wake spontaneously in the pre-dawn hours. Some of these families carry mutations affecting key clock genes; one of the earliest well-characterized examples involved changes in *PER2*. The details are molecular and intricate, but the human picture is simple: an internal night that arrives early and ends early, regardless of dinner invitations or late-night television. The mirror image exists as well: delayed sleep phase tendencies that make sleep onset drift late, often to the distress of people trying to fit into conventional work or school schedules.

These rare families are not the whole story, but they matter because they demonstrate causality. They show that altering clock machinery can alter sleep timing. That is difficult to square with the idea that chronotype is merely a habit, like preferring tea to coffee.

For most people, the genetics are not single-gene destiny but polygenic nudges. Many small differences in clock-related genes, light sensitivity, and sleep homeostatic traits can accumulate into a noticeable phenotype. It is more like height than eye color: strongly influenced by biology, expressed across a spectrum, and shaped by environment in ways that can mislead casual observers. A tall person can slouch, a short person can wear heels, but the underlying range remains.

This is where the term "social jetlag" earns its bite. The phrase refers to the discrepancy between an individual's internal time and the external schedule imposed by work, school, or social obligations. Many people live on two time zones: one from Monday to Friday, another on Saturday and Sunday. They may wake at 6:30 a.m. for work and at 10 a.m. on free days, then lie awake late Sunday night because their circadian phase has been pushed later by weekend sleep and evening light. On Monday, they feel as though they have flown across an ocean, despite never leaving their bedroom. The body is not being melodramatic; it is doing time zone arithmetic without the plane.

Social jetlag is often discussed as a lifestyle issue, but it has deeper implications for health and for fairness. If a schedule systematically forces a subset of people to operate during their biological night, then performance, mood, and metabolic regulation are being asked to function in a low-power state. The cost is not always dramatic enough to be diagnosed as a disorder. It can be a quiet tax: more caffeine, more irritability, more errors, more comfort eating at odd hours, more reliance on willpower to do what biology would otherwise handle automatically.

The two-process model helps disentangle what is being misaligned. In a late chronotype, the circadian wake-maintenance zone extends into the late evening, keeping the sleep gate closed. That person can accumulate very high homeostatic pressure and still find sleep elusive at 10 p.m. If they must wake at 6 a.m., they are cutting sleep short not just in quantity but in circadian placement. They are rising when the circadian system has not yet delivered its full morning alerting push. The homeostat (Process S) may be partially discharged, but not enough; the circadian drive (Process C) is also poorly timed for the demanded wake period. In a chronic state, the person may no longer feel acutely sleepy at bedtime, because the circadian signal is still buoyant, yet feel profoundly unrefreshed in the morning, because both processes are working against them.

The lived experience of chronotype is therefore not merely "I like staying up." It is "my sleep gate opens later, and my morning lift arrives later." Those are not preferences; they are timings.

If the story ended there, chronotype would look like a smooth dial that turns with genes and light. The brain, however, has another feature that makes sleep timing feel hard-edged: the circuitry that flips states.

In the hypothalamus, sleep-promoting neurons in the ventrolateral preoptic area (VLPO) and wake-promoting arousal systems inhibit each other in a mutually antagonistic arrangement, sometimes described as a flip-flop switch. The metaphor is apt: a flip-flop is designed to be stable in either of two states and to switch quickly rather than hover in a noisy middle. This architecture helps explain why sleep onset can feel sudden when it finally arrives, and why partial arousal can snap a person awake. It also helps explain why the sleep gate, once open, can feel like a genuine portal rather than a gentle slope. When the circadian system withdraws wake drive and the homeostatic pressure is high enough, the balance tips; the VLPO gains the upper hand, and wake-promoting systems are actively suppressed. When the balance shifts back, wakefulness is likewise stabilized.

A flip-flop circuit is a good solution to a biological problem: animals that drift in and out of sleep unpredictably are vulnerable. Yet it means that small shifts in timing signals can produce large subjective consequences. Move the circadian gate by an hour, and you may not get an hour's later sleep onset; you may get a qualitatively different evening, with more restless time in bed, more temptation to compensate with screens and snacks, and more anxiety about the coming day. The edges matter.

Chronotype also interacts with cultural assumptions in a way that can distort self-understanding. A late chronotype forced into early mornings often appears to have insomnia, because they cannot fall asleep "on time." But if their internal dusk arrives later, the problem may not be an inability to sleep; it may be an attempt to sleep before the system will permit it. Conversely, an early chronotype living in a late-night culture may interpret their consistent early waking as a disorder, when it may be their circadian system doing exactly what it has been tuned to do.

Physiological markers help separate these possibilities from guesswork. One widely used marker is dim light melatonin onset (DLMO), the time in the evening when melatonin levels begin to rise under dim light conditions. Melatonin is not the cause of sleep by itself, but its onset is a reliable flag that the circadian system has entered its biological night. Measuring DLMO can reveal whether a person who struggles to fall asleep at 10 p.m. has an internal dusk at 8 p.m. or at midnight. That difference is not semantic; it changes what is likely to help. Bright light timed in the morning, dim light in the evening, and consistent wake times can shift circadian phase, but the direction and magnitude depend on where the person's clock currently sits.

None of this should be taken to mean that habit is irrelevant. The circadian system is plastic. People can entrain to schedules, and the homeostatic drive is exquisitely sensitive to how long one stays awake, how much one naps, and how stimulants are used. A person can push themselves later with bright evenings, irregular sleep, and weekend drift. They can pull themselves earlier with steady mornings and protected nights. Yet the ease with which these changes occur varies by person, and that variability is part of the phenotype. Some people shift readily; others resist. Some pay a high price in mood and performance during shifts; others feel only mild discomfort. Even the response to a late-night coffee differs: caffeine's blockade of adenosine is one lever on Process S, but its impact depends on the timing of Process C and on how strongly the person's sleep homeostat responds.

An image helps keep the balance between biology and behavior honest: chronotype is like living near a shoreline. You can rearrange your furniture, change your routine, and even build a seawall, but you cannot negotiate with the tide schedule as if it were a personal whim. At the same time, you can choose whether to open the windows, whether to install blackout curtains, and whether to keep a bright lamp blazing at midnight. The environment you create feeds back into the clock.

The social consequences of chronotype become clearest when two people share a household or a workplace. One falls asleep at 9:30 p.m. and wakes at 5:30 a.m. The other cannot fall asleep before 1 a.m. and does their best thinking at midnight. If both are forced into a 7 a.m. commute, the late chronotype is chronically running a deficit; if both are pushed into late-night deadlines, the early chronotype is. People often treat these mismatches as incompatibilities of character, when they are often incompatibilities of phase.

A small but telling example comes from medicine itself. Medical training has long valued endurance and early starts, sometimes as a badge of seriousness. Yet a resident on a 4 a.m. rotation who is also a late chronotype is not merely "unmotivated" if they struggle;

they are being asked to perform complex cognitive work during their biological night, with high homeostatic pressure and low circadian support. Another resident might find the same schedule unpleasant but manageable. The difference can look like grit, when it is partly timing.

Recognizing chronotype as biology does not solve the practical problem of shared schedules, but it changes what counts as a reasonable question. The question stops being "Why can't you just go to bed earlier?" and becomes "What time does your system make sleep easy, and what can we change—light, timing, obligations—to reduce the mismatch?" Even small adjustments can matter: later school start times for adolescents, flexible work hours, or simply protecting morning light and evening darkness as if they were nutritional inputs rather than aesthetic choices.

Chronotype is sometimes described as an internal preference, but it is better understood as an internal prediction. The circadian system is constantly forecasting when the environment will be safe and useful: when food is likely, when predators are likely, when social activity is likely, when darkness will make vision poor. In modern life, the forecast is often asked to accommodate unnatural lighting and irregular schedules. Some clocks bend; some resist. The resulting chronotypes are not quirks at the edges of society. They are the lived expression of a distributed timing system trying to keep a body coherent.

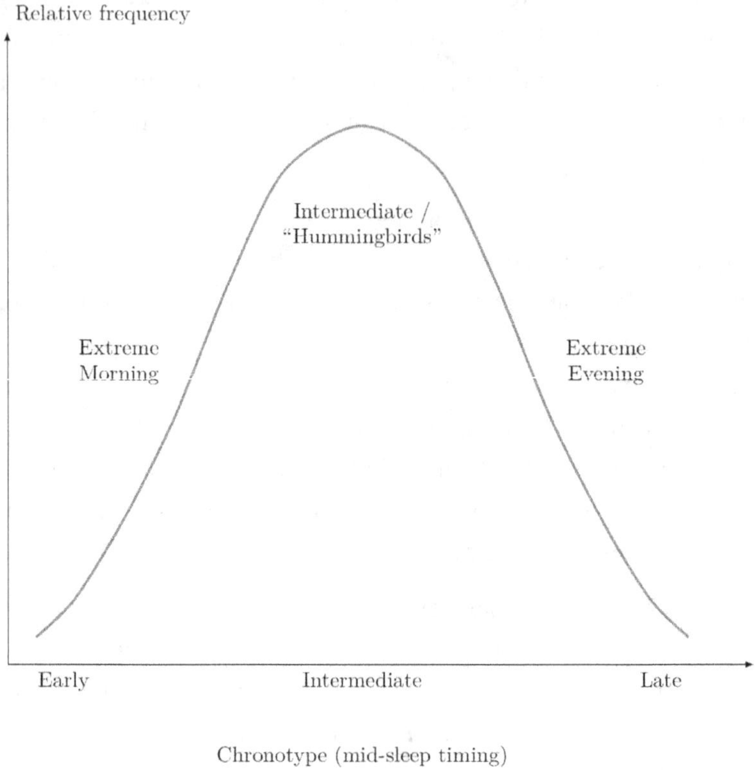

The distribution curve is a reminder that most people fall somewhere in the broad middle, able to tolerate moderate shifts, while a substantial minority cluster toward the tails where the mismatch with conventional schedules becomes chronic. Those tails are not errors. They are variations that likely persisted because different timings can be useful in a group, spreading vigilance and labor across the day and night.

The most humane lesson of chronotype is also the most practical: sleep timing has a biological contour. You can fight it, but you pay. You can respect it, but you may need tools: morning light as a phase-setter, evening darkness as a signal, and routines that stop asking the flip-flop circuitry to hover in an unstable middle. When the horizon of sleep is treated as a real boundary rather than a personal failing, the daily negotiations around bedtimes and alarm clocks become less moralistic and more accurate, and the body's stubbornness starts to look like a form of integrity.

5.3 The Chemical Twilight

A light can be turned off in an instant, but dusk takes its time. Even in a room with no windows, the body tends to stage its own twilight: appetite shifts, skin temperature changes, eyelids grow heavy, thoughts lose their sharp edges, and the same problems that seemed solvable at 4 p.m. begin to feel oddly permanent by 11 p.m. It is tempting to treat that drift as purely psychological, a softening of attention. Yet much of it is chemical and timed, carried by signals that behave less like a gentle mood and more like a set of appointments kept with surprising punctuality.

Two hormones anchor this daily choreography from opposite ends of the night. Melatonin is the best-known messenger of biological dusk, rising in the evening and peaking during the dark phase. Cortisol, often branded the "stress hormone," surges toward morning and helps prime the body for waking life. Neither is a simple on/off switch. Both are better understood as public announcements from the circadian system to the rest of the body, messages that coordinate organs that cannot all "look" at light directly. Together they translate an internal clock into a whole-body schedule, and they do it in a way that interacts intimately with the sleep gate and with chronotype.

Melatonin has an origin story that fits its cultural reputation as a sleep potion, even though the truth is more interesting. The hormone is secreted by the pineal gland, a small structure near the center of the brain that has been mystifying anatomists for centuries. René Descartes, in the seventeenth century, famously speculated that the pineal might be the "seat of the soul," partly because it is singular rather than paired. That idea is no longer in circulation, but the pineal's role is still oddly poetic: it produces a chemical signal of darkness, a hormone that tells the body, in effect, "night has begun."

The pineal gland does not decide this on its own. It is tightly controlled by the brain's central circadian pacemaker in the suprachiasmatic nucleus (SCN), a tiny region in the hypothalamus. When light enters the eyes, specialized retinal cells convey that information to the SCN, which then influences pineal output through a multisynaptic pathway involving autonomic circuits. In everyday terms, the pineal is an obedient servant of the clock, and the clock is an obedient student of light.

The most useful concept here is not melatonin as "sleep hormone," but melatonin as *time stamp*. Under dim light conditions, melatonin begins to rise in the evening at a fairly consistent internal time. This onset is called *Dim Light Melatonin Onset* (DLMO), and it is often treated as the body's true dusk. The "dim light" part matters because bright light can suppress melatonin, masking the underlying clock time. When measured carefully, DLMO is valuable precisely because it tends to be less confounded by activity, posture, meals, and sleep itself than many other rhythms. It provides a crisp physiological marker in a system that otherwise hides behind daily behavior.

DLMO also helps explain why melatonin supplements sometimes feel like magic and sometimes feel like nothing at all. If melatonin's role is to announce biological night, then taking it at the right time can help nudge the system earlier, especially for someone whose internal dusk is delayed. Taken at the wrong time, it may be sedating without being helpful, or simply irrelevant. The clock cares not only about the presence of a signal but about its timing.

Melatonin's daily curve also has an interpretive trap. People assume that because melatonin rises at night, it *causes* sleep. In reality, melatonin is more like a flag that the circadian system raises when it believes darkness has arrived. It can promote sleepiness and facilitate sleep onset in part by acting on receptors in the SCN and elsewhere, but it is not sufficient to produce sleep in a wide-awake brain with low homeostatic pressure. This is one reason a melatonin tablet cannot reliably cancel a late-afternoon nap or a long lie-in. The homeostatic drive still matters. The sleep gate still matters. Melatonin is one of the clock's tools for opening that gate, not a crowbar that can pry it open whenever we wish.

If melatonin is dusk, cortisol is closer to dawn. Cortisol is produced by the adrenal cortex under control of the hypothalamic—pituitary—adrenal (HPA) axis, a system famous for its role in stress responses. But cortisol is not only a crisis hormone; it is also a daily metabolic coordinator. In healthy individuals, cortisol follows a robust circadian rhythm, with low levels in the evening and early night and a rise toward morning. Many people experience a pronounced increase shortly after waking called the *Cortisol Awakening Response* (CAR). This rise helps mobilize energy, influences immune function, and can sharpen attention. It is part of why waking up can feel like a biological event rather than a simple decision. Even before the eyes open, the body is shifting gears.

Cortisol's reputation suffers because it is implicated in anxiety, chronic stress, and many disease processes. Yet in the context of daily timing, cortisol is neither villain nor hero. It is a signal that promotes wakefulness and readiness, and its timing matters. A cortisol rhythm that is mistimed or flattened can leave a person feeling as though their mornings never truly begin or their evenings never truly end. In shift work and in chronic circadian misalignment, cortisol patterns can become blunted or shifted, and the subjective experience is often a persistent sense of being out of phase with the day.

These hormonal rhythms are not isolated curves; they sit on top of, and interact with, the two forces that determine sleep timing. The homeostatic drive builds with time awake and is strongly influenced by sleep loss, while the circadian system gates sleep and wake by changing thresholds and arousal signals. Melatonin and cortisol are among the chemical ways the circadian system broadcasts its current state to the rest of the body, aligning sleep propensity, temperature regulation, metabolism, and alertness.

The phrase "chemical twilight" deserves to include a third actor that is not a classical hormone but behaves like one in the story of sleep pressure: adenosine. Adenosine is produced throughout the body, but within the brain it acts as a neuromodulator that can accumulate with prolonged wakefulness. One influential line of work has shown that extracellular adenosine in the basal forebrain tends to rise during wake and sustained wakefulness and then fall during recovery sleep. The basal forebrain is not a sleepy backwater; it is a hub of arousal and attention. When adenosine levels rise there, they can dampen wake-promoting neurons through receptors such as A1, making wakefulness harder to sustain. This is not mere metaphorical "fatigue"; it is a change in the chemical environment of circuits that keep the cortex engaged with the world.

Caffeine's everyday effectiveness is a clue. Caffeine does not replace sleep. It mainly blocks adenosine receptors, preventing the brain from reading its own sleep pressure accurately. In the language of Process S, caffeine does not empty the tank; it covers the gauge. That can be useful, even life-saving in some contexts, but it also means the homeostatic debt can keep accumulating in the background. When caffeine wears off, the pressure has not evaporated. It has merely been postponed.

Adenosine also shows why small chemical changes can produce abrupt state changes. The sleep—wake system is not a dimmer switch; it is closer to a bistable circuit. Sleep-promoting neurons in the VLPO and wake-promoting systems inhibit each other, creating a flip-flop-like architecture that supports stable wakefulness and stable sleep with relatively rapid transitions. In such a system, neuromodulators like adenosine can act as gentle fingers on the scale. They need not be overwhelming to be decisive. A modest increase in sleep-promoting tone, or a modest decrease in arousal tone, can tip the balance.

Another chemical actor with a quieter reputation is prostaglandin D_2 (PGD_2). Prostaglandins are lipid-derived signaling molecules involved in inflammation and many physiological processes. PGD_2 has been described as a potent sleep-promoting factor, and one proposed pathway links it to adenosine: PGD_2 can stimulate adenosine release, and adenosine signaling can activate sleep-promoting neurons while suppressing arousal systems, including histaminergic pathways. The details are complex, but the conceptual payoff is simple and intuitive: immune signaling and metabolic signaling can intersect with sleep regulation. The body's chemistry does not respect the neat boundaries between "sleep," "stress," and "inflammation" that appear in self-help categories.

A concrete everyday example shows these intersections: the peculiar sleepiness of illness. When people develop a viral infection, many feel a heavy, irresistible sleep drive that does not resemble ordinary fatigue. Part of that may be behavioral and motivational, but part reflects immune and inflammatory signals that interact with sleep-promoting chemistry. The body seems to choose sleep as a strategy, not merely as a consequence of weakness. Conversely, in some inflammatory states sleep can be fragmented and unrefreshing, suggesting that the same signaling families can destabilize the flip-flop rather than smoothly promoting restorative sleep. The chemistry is not a single lever but a set of interacting knobs.

A historical anecdote from circadian research adds a useful cautionary note about interpreting hormones. When researchers first began to use physiological rhythms to infer circadian phase, they ran into a pervasive problem called *masking*. Many variables change with the clock, but they also change with behavior. Body temperature is strongly circadian, but it is also influenced by exercise, meals, and ambient conditions. Cortisol responds to stress, but stress is not evenly distributed across the day. Even sleep itself masks many rhythms because it changes posture, breathing, and sensory input. Lewy and Sack's work in the late 1980s helped establish DLMO as a practical marker partly because melatonin under dim light is comparatively less vulnerable to these behavioral overlays. The moral is not that melatonin is pure and cortisol is noisy. The moral is that physiology has layers: what the clock is doing, what behavior is doing, and what the environment is doing. The art of chronobiology is learning how to separate them.

The concept of DLMO also reframes what "night owl" means in chemical terms. A late chronotype is not merely someone who enjoys midnight. It is often someone whose melatonin onset occurs later relative to clock time. Their biological dusk may begin at 10 p.m. when a colleague's begins at 8 p.m. Their cortisol rhythm may also be shifted, with the morning rise occurring later. If both people are forced into a 6:30 a.m. commute, one is commuting in biological morning, the other in biological night. They may look equally awake from across the room, but their chemistry is not matched.

This has consequences for mood and cognition that are easy to misread. The late chronotype who must wake early may feel anxious at night, not because they are psychologically troubled, but because they are trying to sleep before their melatonin rise and before circadian wake promotion has relaxed. The early chronotype forced to stay up late may feel suddenly flat or emotionally brittle, not because they lack social stamina, but because cortisol is low, melatonin may be rising, and arousal circuits are losing their chemical support. Social judgments pile quickly onto these moments. Biology rarely gets the benefit of the doubt.

There is also a deeper point about the body's distributed timekeepers. Melatonin and cortisol do not merely mark time; they help *coordinate* time across tissues. Almost every organ has its own clock-like gene expression rhythms. The liver anticipates meals. Fat tissue anticipates fasting and feeding. Immune cells anticipate diurnal patterns in exposure and activity. The SCN acts as a conductor, but hormones are among the signals that carry the beat to distant players. Melatonin receptors exist in many tissues; cortisol acts widely through glucocorticoid receptors, influencing gene expression. When these hormonal rhythms are mistimed, organs can drift out of alignment with each other. A person may feel this as vague malaise: hunger at odd hours, sluggish digestion, restless sleep, a sense that the body is conducting multiple days at once.

To appreciate how sharply timed these signals can be, consider a traveler flying east across several time zones. On arrival, the local dusk arrives, streetlights turn on, restaurants fill. But the traveler's DLMO may not arrive for hours; melatonin may remain low, and the circadian system may still be broadcasting "day." They may feel alert at local midnight. Then, in the local morning meeting, cortisol may not yet have risen in their internal schedule; the chemical dawn is delayed. They experience jet lag not as an abstraction but as a mismatch between the world's light-dark cycle and the body's hormone schedule. The mismatch resolves only when light exposure and behavior gradually shift the internal clock, pulling DLMO earlier and shifting the cortisol rhythm toward local time. The body is not confused; it is conservative, refusing to rewrite its schedule on the basis of a single night.

Another example, more modern and more intimate, is the bright screen at night. Many people have experienced the strange effect of feeling sleepy, checking a phone, and then feeling awake again. Part of this is psychological stimulation. Part is the suppression or delay of melatonin by light, especially short-wavelength light, which signals "day" to the circadian system. The effect is not necessarily dramatic in everyone, but for a vulnerable chronotype it can matter. A delayed melatonin rise delays the sense of biological dusk, which keeps the sleep gate from feeling fully open. The person then compensates by staying up longer, increasing homeostatic pressure, and may eventually fall asleep from sheer accumulation, but at a cost: shorter sleep, later wake times on free days, and greater social jetlag.

Chemical timing also helps interpret the popular idea of "sleep quality." People often judge sleep by how quickly they fall asleep or how long they sleep. Yet the timing relative to melatonin and cortisol rhythms matters. Sleeping at a circadian phase that the body does not "expect" can produce lighter, more fragmented sleep, not because the person is failing at relaxation but because arousal systems are being maintained by circadian signals and because hormonal rhythms are not aligned with the sleep episode. Forced desynchrony studies have shown that sleep propensity and certain features of sleep architecture vary with circadian phase independent of time awake. In ordinary life, this means that a daytime sleep episode after a night shift is often fighting the biology of day.

The chemical twilight is not only about sleep onset. It is also about sleep's internal texture. Melatonin rises, core body temperature tends to fall, and the brain's arousal systems are gradually inhibited. Within sleep, the early part of the night often carries more slow-wave sleep, reflecting high homeostatic pressure and deep recovery, while later parts of the night often contain more REM sleep, which is strongly shaped by circadian phase. These patterns are not rigid rules, but they reveal an underlying design: the night is structured, not amorphous, and part of that structure is imposed by chemical and circadian timing.

It can be tempting to use this knowledge to chase an ideal schedule with pharmaceutical precision: melatonin at the perfect minute, caffeine only in mathematically safe windows, a dawn simulator tuned to the millilux. There is a place for careful timing, especially for circadian rhythm disorders and for shift workers, and clinical chronobiology has developed practical tools. But there is also a simpler, more humane application: understanding why certain struggles feel stubborn. A person trying to sleep before DLMO is often fighting their own dusk. A person waking during their cortisol nadir is trying to begin a day before their chemical dawn. A person using caffeine late is not adding wakefulness so much as borrowing it against a homeostatic debt that will be collected later.

The body's twilight is chemical because it must be. A clock that stayed trapped in the brain would be useless to the liver, the immune system, the skin, the gut. Hormones are the broadcast medium. Melatonin tells tissues that night is underway; cortisol helps coordinate the return to daytime metabolism and vigilance. Adenosine and related sleep-promoting factors translate cellular activity into sleep pressure, pressing on a flip-flop circuit that can turn a gradual chemical drift into a sharp change of state. The result is a daily transition that feels, from the inside, like a threshold being crossed: the moment when the world has not changed very much, but the body has decided that it is night.

Chapter 6

Energy and Defense

The body faces a fundamental resource trade-off between fueling immediate physical activity and maintaining a standing army against microscopic invaders. By temporally segregating glucose processing from peak inflammatory responses, the circadian clock optimizes our survival odds but creates specific windows of vulnerability. We will discover why a late-night meal lingers dangerously in the bloodstream and why the immune system's vigilance cycles often dictate the precise hour when maladies strike.

6.1 The Metabolic Partition

A spoonful of sugar is never just a spoonful of sugar. The same glucose molecules can be welcomed, efficiently packed away, and burned for work at one time of day, yet handled clumsily at another. Many people discover this only when a continuous glucose monitor traces the body's internal weather: breakfast produces a modest hill, dinner the same size becomes a mountain. The surprising part is not that sleep, stress, or exercise matter. The surprise is that even when those are held steady, the body keeps time. Metabolism is scheduled.

Glucose tolerance is a phrase clinicians use for a simple idea: how well the body keeps blood sugar in a safe range after eating. When glucose tolerance is good, blood sugar rises and then returns to baseline smoothly. When it is poor, the rise is higher and the return slower. Over years, that pattern pressures the pancreas, reshapes the liver, and nudges tissues toward insulin resistance, the hallmark of type 2 diabetes. Glucose tolerance is not constant across the day; in humans it tends to be better early and worse late, even for identical meals.

To understand why, it helps to treat eating as a physiological event with two jobs. The first job is obvious: provide energy and building blocks. The second job is less obvious: coordinate multiple organs so the incoming nutrients do not destabilize the bloodstream. The body has to decide, in minutes, where to send glucose and whether to store it, burn it, or convert it into other forms of chemical currency. That decision depends on insulin, the hormone made by the pancreas's *beta cells*. Insulin is the message that says, "fuel has arrived; open the doors." Muscle and fat respond by pulling glucose out of blood; the liver responds by switching from exporting glucose to storing it as glycogen.

This system looks like plumbing, but it behaves like an orchestra with a conductor. The conductor is not a single clock in the brain issuing commands to passive organs. Each major metabolic organ carries its own timekeeping machinery, a local set of molecular gears built from genes that turn one another on and off over roughly 24 hours. The pancreas is one of those timekeepers. Its clock does not merely keep score; it changes the way beta cells perform, altering how readily they secrete insulin and how well they cope with the stress of repeated surges.

A useful way to picture the daily schedule is as a partition of metabolic labor. During the biological day, when the organism is expected to be active, the system is biased toward handling incoming fuel: higher insulin sensitivity (tissues respond well to insulin), brisk beta-cell responsiveness (the pancreas secretes promptly), and readiness in the liver and muscles to store or use glucose. During the biological night, the same machinery is biased toward a different set of priorities: cellular maintenance, repair, and the careful rationing of stored fuel. That nighttime program is not a moral judgment on late dinners; it is an evolutionary logic. For most of human history, and for most animals today, nighttime was a poor time to forage widely and an excellent time to avoid predators, heal small injuries, and consolidate memory. When food intake becomes common at night, it forces the body to do daytime work during a period it has reserved for backstage tasks.

The pancreas, in particular, is not an all-hours diner. Beta cells are unusually sensitive to the chemical burden that comes with their job. Every time they secrete insulin, they must manufacture, fold, package, and ship a protein hormone at high speed. Protein folding is error-prone; misfolded proteins can stress the endoplasmic reticulum, the cell's internal factory floor. Over time, repeated metabolic overdrive contributes to beta-cell dysfunction. The idea that the pancreatic clock protects the beta cell is therefore intuitive: schedule the most demanding secretory work for the part of the day when the cell's support systems are primed, and reserve the night for cleanup and repair.

A historical glimpse of how central insulin is to this story comes from one of medicine's most dramatic before-and-after moments. In 1921, Frederick Banting and Charles Best began experiments in Toronto that led to the isolation of insulin. Before insulin therapy, a diagnosis of type 1 diabetes, especially in a child, was often a slow emergency. Photographs from that era are hard to forget: children reduced to skin and bone, kept alive briefly by starvation diets that could delay coma but also delayed childhood itself. When insulin arrived in clinics in the early 1920s, it did not merely lower a lab value; it returned appetite, growth, and life. That history can make insulin feel like a simple replacement, a missing substance supplied from outside. Yet in type 2 diabetes the tragedy is subtler: insulin is present, sometimes abundant, but the timing and coordination of its signal are compromised. The system becomes noisy. The body is still trying to follow a schedule, but modern habits keep changing the clock on the wall.

One of the clearest modern habits is eating late. Imagine two identical dinners: same calories, same carbohydrates, same person, same day. One is eaten at 8:00 AM (unusual, but possible), the other at 8:00 PM. Even if the person stays awake in both cases, the glucose response often differs: the evening meal produces a higher peak and a slower decline. Part of that difference comes from insulin sensitivity. Muscle is not equally receptive to insulin at all times; in the late biological day, it tends to require a louder insulin "voice" to open the same glucose "doors." Another part comes from the beta cell itself, which appears less eager to respond as biological night approaches.

There is a further twist that makes the evening especially awkward: melatonin. Melatonin is commonly called the "darkness hormone," secreted by the pineal gland as light fades. It helps coordinate the body's sense of night, nudging physiology toward sleep and signaling peripheral tissues that the external world is dim. Melatonin does not merely float benignly in the bloodstream; it binds to receptors in many tissues, including the pancreas. For some individuals, genetic variation in a melatonin receptor (often discussed under the gene name *MTNR1B*) can magnify the glucose cost of eating late. In that case the body receives two conflicting messages at dinner: nutrients are arriving, yet the hormonal banner overhead reads "night." The metabolic partition becomes a tug-of-war.

The practical consequences show up most starkly in shift work. Modern economies depend on night nurses, emergency physicians, long-haul drivers, police officers, factory workers, and many others whose jobs run against the solar day. Shift work is not merely "staying up late." It can create circadian misalignment: the brain's central clock, calibrated mainly by light, may remain anchored to day-night cycles, while behavior (sleep and meals) shifts to a different schedule. Meanwhile, peripheral clocks in the liver, pancreas, gut, and fat are sensitive not only to light-driven signals but also to food timing. When meals occur at night, peripheral clocks may drift, sometimes out of step with one another. The result is internal desynchrony, like a city in which the public transit, hospitals, and power grid all run on different timetables.

In controlled laboratory studies that simulate night work, a striking pattern emerges: eating during the biological night worsens glucose tolerance, but restricting food to the biological day can protect metabolic control even when the sleep schedule is inverted. That finding carries an important emotional implication. Many shift workers are told, implicitly or explicitly, that their bodies are simply not disciplined enough. In reality, a large part of the struggle is structural. The body's metabolic equipment is being asked to perform its heaviest labor at the time it is programmed for maintenance. Meal timing becomes a lever not because willpower is magical, but because the circadian system is listening.

It is tempting to interpret the daily rhythm of glucose handling as a simple "morning good, evening bad" rule. Real physiology is messier. Exercise can improve evening glucose control because muscle contractions pull glucose into cells through insulin-independent pathways. A large breakfast may overwhelm capacity even in the morning. Sleep deprivation can worsen glucose tolerance at any hour. Yet the rhythm persists across many conditions, which suggests that it is not merely a byproduct of behavior. The body's clocks are actively shaping the playing field.

The pancreas's clock is built from the same core molecular parts found elsewhere: genes whose protein products inhibit their own production in a delayed feedback loop, generating oscillations. When these oscillations are robust, beta cells anticipate daily cycles of demand and adjust their secretory machinery accordingly. When the clock is disrupted—by chronic circadian misalignment, irregular sleep, or metabolic disease itself—beta cells can lose this anticipatory tuning. The point is not that a beta cell checks a wristwatch. The point is that its internal state changes across the day: which genes are expressed, which enzymes are plentiful, how ready calcium signaling is to trigger insulin release, how resilient the cell is to stress. Time is written into the cell's readiness.

A useful analogy is a bakery. The bakery can, in principle, bake at any hour. But if deliveries, staff shifts, oven maintenance, and cleaning are scheduled for certain times, trying to produce the most elaborate pastries during the cleaning window is a recipe for mistakes. Late-night eating is not just "extra work"; it can be work at the wrong phase, competing with necessary restoration.

One reason the late meal can echo into the next day is that glucose control is not a single-meal phenomenon. The liver's glycogen stores, the muscle's insulin sensitivity, the hormones that govern appetite and stress, and the sleep that follows all interact. A very late dinner can keep glucose elevated longer into the night, and elevated glucose often travels with elevated insulin. That pairing can nudge the body toward storing more energy while simultaneously leaving tissues less responsive the next morning, a kind of metabolic jet lag. People sometimes describe this as "my body doesn't reset overnight," which is not far from the truth: there is no reset button, only cycles.

Consider a concrete example. Two coworkers share the same cafeteria. One eats most calories between 8:00 AM and 3:00 PM, with a substantial lunch. The other eats lightly in the day and makes up for it after 8:00 PM, partly because work is busy and partly because dinner is the main family meal. Neither is overeating by intention; both consume similar total calories. Over months, the second person may notice that hunger arrives later and later, that breakfast feels optional, and that energy slumps occur in the mid-morning. If they check blood sugar, they might find higher post-dinner spikes and, eventually, higher fasting glucose in the morning. The body is adapting to a late fuel schedule, but the adaptation is metabolically costly because it pushes against the partition the circadian system prefers.

A deeper evolutionary logic sits underneath these patterns. During the active phase, an organism must be ready to take advantage of unpredictable opportunities: find food, escape danger, negotiate social hierarchies, travel, build. Those tasks require flexible fuel use. During the rest phase, the priorities shift toward processes that are expensive but not urgent in the moment: DNA repair, protein quality control, synaptic remodeling, and immune housekeeping. The metabolic partition is a way of avoiding simultaneous peak demands. If a large glucose load arrives during the rest phase, the body must allocate resources to processing it while still attempting to run the maintenance program. That conflict can produce chronic wear.

This is also why "calories are calories" is an incomplete sentence. Physics is unimpressed by the clock, but biology is not a closed calorimeter. The same calorie has different hormonal and cellular consequences depending on when it arrives. In some people, eating earlier in the day improves insulin sensitivity and lowers blood pressure even without weight loss, suggesting that the body's timing machinery can be nudged back toward a cooperative rhythm. The message is not that everyone must eat breakfast at dawn or fear a late-night snack as if it were poison. The message is that the metabolic system contains a time-based bias, and ignoring it is like swimming against a current: possible, sometimes necessary, but tiring.

A final nuance matters for anyone who has felt blamed by simplistic advice: individual variability is real. Chronotype—whether someone is naturally more alert early or late—influences when the body's internal day begins. Age shifts timing as well; adolescents tend to run later, while older adults often run earlier. Genetics adds another layer, as do medications and metabolic health. Two people can eat at the same clock time while eating at different *biological* times. That is one reason a universal rule is hard to write. Yet the broad human pattern remains: glucose tolerance tends to be higher earlier in the biological day, and circadian misalignment tends to worsen metabolic control beyond the effect of sleep loss alone.

The most compassionate way to think about the metabolic partition is as a design constraint. It is a feature that helped a diurnal primate thrive in a world where darkness meant limited food and higher risk. In the modern world, the constraint has not disappeared; it has become more visible. Electric light, round-the-clock commerce, and abundant calories have created a new ecological niche: the well-fed night. The pancreas did not evolve for this niche, and it responds the way an ancient system responds when its schedule is repeatedly overridden: it strains.

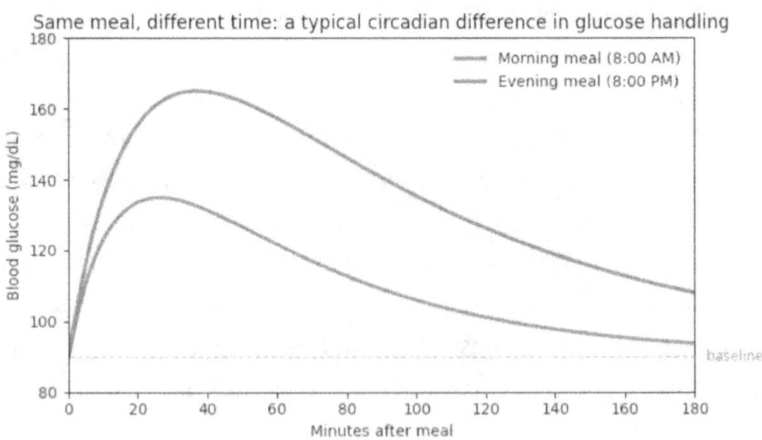

If the graph feels discouraging, it can also be read as hopeful. A rhythm implies a handle. People cannot easily rewrite their genes, and many cannot change work schedules, family obligations, or the realities of shift-based jobs. But timing is a behavioral signal that the body's clocks understand. For someone at risk of diabetes, even modest shifts—bringing dinner earlier when possible, avoiding very large meals close to bedtime, placing the most carbohydrate-heavy meal earlier in the waking interval, or pairing late eating with physical activity—can reduce the size and duration of glucose excursions. These are not cosmetic changes. Each reduction in peak and each shortening of time spent above baseline is a small decrease in the chemical stress that accumulates across years.

One can see the metabolic partition, then, not as a rule carved into stone but as a negotiation between ancient biology and modern life. The circadian system is not trying to punish late eaters; it is trying to keep multiple organs from talking over one another. When meals align with the active phase, the conversation becomes clearer: insulin speaks, tissues listen, glucose settles. When meals arrive at biological night, the dialogue grows strained, and the bloodstream carries the record of that strain in the form of a higher peak and a longer descent.

6.2 Gated Immunity

The immune system is often described as an army, but it behaves more like a city's emergency services. Patrols circulate, specialized teams stage near likely trouble spots, and information is routed to central hubs where decisions are made about escalation. Like any emergency system, it faces a dilemma: vigilance is expensive. An immune response burns energy, diverts nutrients, and risks collateral damage to healthy tissue. Fever is a controlled fire; inflammation is a controlled flood. Too little, and infections gain ground. Too much, and the body injures itself.

Daily timekeeping offers a way out of that dilemma. Rather than holding every defense mechanism at full readiness all the time, the body schedules parts of immunity the way it schedules metabolism: heightened watchfulness when threats are most likely, and quieter administrative work when the organism is supposed to be resting. The resulting pattern can be called *gated immunity*. "Gated" does not mean locked shut; it means that the immune system opens and closes certain doors in a time-dependent way. Cells traffic differently, tissues display different "welcome signs," and inflammatory signals are allowed to rise or are actively restrained depending on circadian phase.

This idea can feel counterintuitive because microbes do not check the clock. A virus can enter at noon or midnight. Yet the immune system does not need perfect prediction to benefit from rhythm. It needs only a statistical advantage: if most exposures happen during wakefulness, when hands touch faces and lungs inhale crowded air, it makes sense to concentrate some defenses at the body's borders during those hours. In many species, including humans, the active phase is also when small injuries occur. A scratch on the skin, a nick in the mouth, a bit of dust deep in the airway are everyday events. The immune system can treat wakefulness as a period of expected minor breaches, and sleep as a period for processing reports, repairing damage, and refining memory.

The notion that immunity has a daily rhythm is older than the molecular clock. Long before genes were measured, physicians noticed that symptoms seemed to "keep appointments." Fevers often climb in the evening. Certain pains are worse at dawn. Even the counts of different blood cells, measured in early clinical laboratories, showed daily swings. For a while these observations were treated as clinical curiosities: the body gets tired, hormones fluctuate, patients behave differently. The more startling claim is stronger: immune rhythms are not merely consequences of daily life; they are actively produced by clocks distributed through the body, including within immune cells themselves.

To make that concrete, it helps to name the players. "White blood cells" is an umbrella term, like "vehicles." Some are fast responders of the innate immune system (neutrophils, monocytes, many macrophages), built for rapid, nonspecific defense. Others belong to the adaptive immune system (T cells and B cells), slower to mobilize but capable of learning and remembering a specific pathogen for decades. Still others (dendritic cells) act as translators, carrying information from the periphery to the adaptive system. All of these cells travel. They spend time in the bloodstream, but blood is not their home; it is their highway.

Their destinations are telling. Barrier tissues—skin, lungs, and the gut—are the front doors of the body. They are where the outside world presses in. Lymph nodes are more like command centers and archives. They are where immune cells meet, exchange information, and decide what kind of response is needed. A dendritic cell that has sampled microbes in the airway can travel to a lymph node and present pieces of those microbes to T cells, initiating a tailored response. In broad strokes, then, one can imagine a daily commute: patrols to the borders during the active period; return to the nodes during rest to file reports and plan.

The "commute" metaphor is not just poetry. It is built into the mechanics of how leukocytes (a general term for white blood cells) move. To exit the bloodstream into a tissue, a leukocyte must slow down and adhere to the vessel wall. That adherence depends on molecules displayed by endothelial cells, the cells lining blood vessels. Endothelial surfaces can become sticky or slick depending on which adhesion molecules they express. Chemokines (chemical attractants) also matter, acting like scent trails that guide cells toward particular tissues. Many of these signals vary with time of day.

One reason they vary is that immune cells have their own clocks. The same core molecular machinery that creates circadian rhythms elsewhere also operates inside macrophages, neutrophils, and lymphocytes. Inside these cells, clock-controlled genes tune readiness: how strongly inflammatory programs can be switched on, how energetically cells can migrate, how avidly they engulf debris. Clock genes do not order an immune cell to "be inflamed at 3 a.m."; they bias probabilities and thresholds. A cytokine signal that would trigger a strong response at one time might trigger a muted response at another because the molecular context has changed.

Another reason is that immune cells are embedded in a body whose nervous and hormonal systems are also rhythmic. Sympathetic nerves, the fast-acting branch of the autonomic nervous system, innervate blood vessels and lymphoid tissues. Their activity changes over the day—night cycle, altering local chemistry. Hormones such as cortisol (a glucocorticoid) rise and fall in predictable patterns, with broad immunomodulatory effects. When these systemic rhythms intersect with immune-cell intrinsic clocks, the result is a coordinated choreography: the same immune cell may receive different instructions depending on the hour.

A historical anecdote illustrates how deeply timing is woven into immunity, even if it was not recognized as circadian at first. Smallpox vaccination, pioneered by Edward Jenner in 1796, spread through a world without antibiotics and with limited understanding of microbes. Clinicians noticed enormous variability: some people mounted robust reactions, others barely "took." The reasons were many—vaccine quality, prior exposure, health, age—but the deeper principle was that immune responses are not fixed outputs. They depend on the internal state of the host at the moment of challenge. Modern chronobiology adds a layer to that older lesson: time of day is part of "internal state." The same antigen encountered at different circadian phases can be processed and responded to differently.

The concept of gating becomes clearest when you imagine two distinct tasks that cannot safely be maximized simultaneously. One task is *frontline interception*: sending cells into barrier tissues where they can rapidly confront invaders. The other is *information processing*: collecting antigen, teaching lymphocytes, pruning reactions that are too broad, and consolidating immunological memory. The first task risks tissue damage because it brings inflammatory machinery into delicate interfaces like the lung. The second task risks delay because it keeps cells in hubs rather than at the borders. A time-based schedule allows the system to lean one way and then the other.

A practical example comes from the mundane experience of seasonal respiratory infections. A person spends the day in shared indoor air: classrooms, buses, offices, shops. Inhaled particles land on the airway epithelium, the cellular lining of the nose and lungs. That epithelium is not passive; it produces antimicrobial peptides and calls for help when irritated. If leukocytes are more likely to be stationed near airway tissues during the active phase, a viral inoculum may be met quickly by resident macrophages and recruited neutrophils. The goal is not necessarily to prevent infection entirely, which can be impossible, but to keep the initial viral expansion small. A smaller initial expansion can change the whole arc of illness: fewer infected cells, less inflammatory spillover, and a lower chance of the immune response itself causing major symptoms.

Yet the story is not simply *"more immune activity is better."* The immune system must also protect the body from its own defenses. Inflammation helps clear pathogens, but it can scar lungs and inflame blood vessels. One reason immune gating likely evolved is to reduce the chance that a vigorous inflammatory response coincides with periods when tissues are doing other essential work, such as the nocturnal repair of DNA damage and the clearance of cellular debris. Sleep is also a period when breathing patterns change and airway clearance is different; flooding the lungs with inflammatory cells at the wrong time could be especially disruptive. Gating is a compromise between protection and preservation.

Mechanistically, one of the most compelling findings in modern circadian immunology is that time-of-day differences are not merely correlations. When researchers disrupt the molecular clock in immune cells, the daily rhythm of certain immune responses can flatten or vanish. That is a strong form of evidence because it shows causality: the clock is part of the machinery, not merely a spectator. Likewise, when neural signals that carry time-of-day information are altered, leukocyte trafficking patterns change. The body's timekeeping network reaches into immunity by more than one route.

The distributed nature of the control matters because it prevents a simplistic picture. There is not a single immune clock that rings like a bell. Instead, there are at least three intertwined influences.

First, immune cells carry intrinsic oscillations that affect their behavior. A macrophage at one circadian phase may be more prone to produce inflammatory cytokines, and at another more prone to engage in tissue-repair programs. Cytokines are the immune system's short messages: proteins such as interleukins and tumor necrosis factor that tell nearby cells to activate, recruit, or calm down. These messages are potent; their timing matters.

Second, endothelial cells and tissue stromal cells (the supportive cells within organs) change their expression of adhesion molecules and chemokines across the day. This is crucial because leukocytes cannot simply decide to exit the bloodstream; they require a receptive vascular surface. If the endothelium in skin and lung is stickier during the active phase, more cells can park there. If it is less sticky during rest, cells tend to remain in circulation or return to lymphoid compartments.

Third, systemic signals provide an overarching rhythm. Hormones and autonomic nerves can tune both immune cells and the tissues they enter. Sympathetic signaling, for example, can modulate vascular tone and the molecular "address labels" that determine where immune cells home. These signals are also sensitive to behavior: stress, exercise, and meals can all modulate them. That means gating is robust but not rigid; it can be overridden in emergencies.

The cost of overriding becomes visible in chronic circadian disruption. Shift work, jet lag, and irregular sleep schedules do not merely make people feel tired; they can disturb immune rhythms. Epidemiological studies have long suggested that shift workers face increased risks of certain infections and inflammatory conditions, though disentangling work, sleep, stress, and socioeconomic factors is challenging. Laboratory work clarifies at least one piece of the puzzle: when the internal timing signals to immune cells and tissues become inconsistent, the trafficking schedule can become less efficient. Cells may be in the wrong place at the wrong time, and inflammatory signals may rise when restraint would be safer or fall when vigilance would be useful.

This inefficiency can be imagined in very physical terms. A neutrophil in the bloodstream is not doing nothing; it is consuming energy and aging. Neutrophils have short lives and are replaced in enormous numbers. Directing them into tissues at a time when they are less needed wastes resources and raises the risk of bystander damage. Conversely, failing to stage sufficient sentinels at barriers during high-exposure periods increases the chance that pathogens establish footholds. Gating is therefore not only about defense; it is about budgeting.

The social dimension is hard to ignore. Many of the people whose work keeps hospitals, logistics, and public safety running are also those who bear the physiological cost of circadian disruption. When immune gating is disrupted, the consequences can show up as more frequent minor illnesses, slower recovery, or heightened inflammation that aggravates chronic disease. These are not dramatic, single-event failures. They are small erosions in resilience that accumulate quietly.

A further nuance, often missed in popular discussions, is that the immune system's rhythms differ across species and even across cell types. In nocturnal animals like mice, which are commonly used in immunology laboratories, the active phase occurs at night. A leukocyte trafficking pattern that peaks during the mouse's resting phase may correspond to daytime in the human, but the biological meaning is "rest" versus "active," not simply "light" versus "dark." This is one reason translation from animal studies must be done thoughtfully. It is not enough to know the clock time; one must know the biological phase.

That caveat does not weaken the central point. It strengthens it by emphasizing that immune rhythms are functional. They are aligned to behavior, to exposure risk, and to energetic constraints. The immune system is not merely rhythmic because everything in biology is rhythmic; it is rhythmic because timing improves the trade-offs.

The daily commuting pattern of leukocytes can be summarized without drowning in detail: during the active phase, more surveillance at the borders; during the rest phase, more congregation in lymphoid tissues for coordination. The exact timing and the exact cell types involved vary, but the logic recurs. The system uses time as an organizing principle.

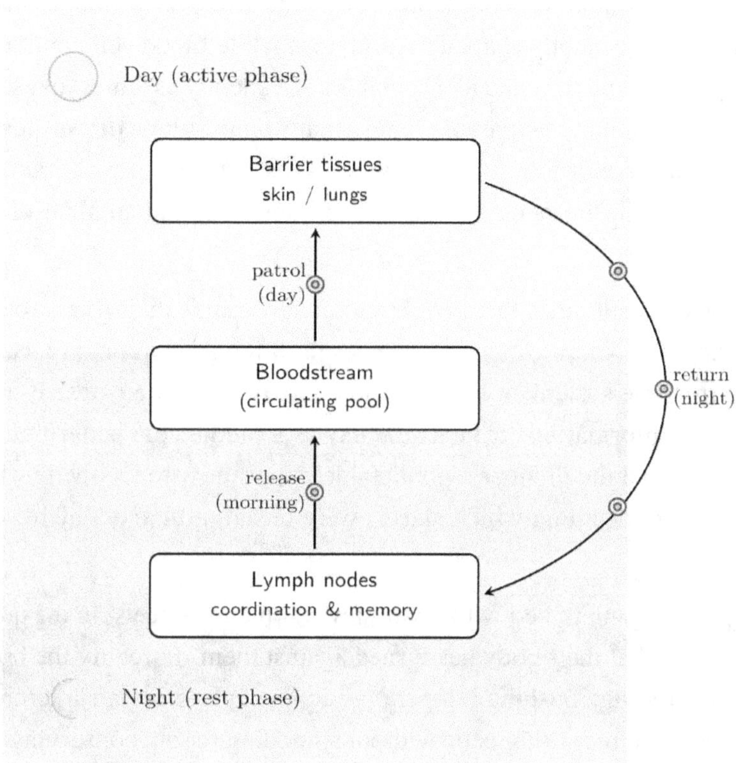

The diagram hides a great deal of molecular detail, but that is part of its honesty. The key point is location. Immune defense is not only about how fierce cells are; it is about where they are stationed when the breach occurs. A perfectly competent firefighter is useless if posted miles away without a road map.

The same location principle helps explain why the immune system can feel different at different times. A vaccine that relies on antigen presentation and lymphocyte activation may be more effective if administered when trafficking and signaling favor strong adaptive responses. An inflammatory insult may be more damaging if it occurs when cytokine brakes are looser. Even the interpretation of clinical blood tests is subtly shaped by timing: a white blood cell count can vary across the day, and a "normal" range silently assumes that samples are taken at comparable times. In hospitals, blood draws cluster in the early morning for practical reasons, which means clinicians often see the immune system at a particular circadian slice of its behavior.

None of this implies that people should reorganize their lives around leukocyte commutes. It does suggest a different kind of respect for the immune system: it is not merely reactive. It anticipates. It uses time as information. It treats the day as a predictable pattern of exposures and the night as a predictable opportunity to reconcile what happened, deciding which alarms were meaningful and which were noise.

A person lying in bed with a mild fever sometimes feels, in the quiet of night, that their body has turned against them. In reality the body is negotiating: pushing temperature up enough to impair microbial replication, recruiting cells without shredding tissue, conserving energy while still paying for defense. The immune system's daily gates do not guarantee safety, but they reduce unnecessary expense. They are the difference between a city with sirens blaring constantly and a city that has learned when to position its ambulances near the stadium and when to bring them back to base to restock, refuel, and read yesterday's reports.

6.3 The Timing of Maladies

A curious pattern emerges when people begin to keep diaries of symptoms. The entries often read less like random misfortune and more like a timetable: wheeze around 4 a.m.; stiffness on waking; headaches that prefer the late afternoon; a heart that seems most likely to betray its owner in the first hours of the working day. Medicine has long been alert to what it calls *diurnal variation*—changes across the day—but the temptation is to treat these regularities as superficial. Perhaps night air is cold. Perhaps morning is stressful. Perhaps people notice symptoms more when they are still. Those explanations sometimes help, yet they miss a deeper point: many diseases are not merely present in the body; they are *modulated* by the body's internal clocks. Symptoms can be read as the visible surface of rhythms in hormones, immune mediators, airway tone, blood pressure, and the very traffic of immune cells.

The most revealing case is inflammation. Inflammation is a word used for heat, swelling, and pain, but biologically it is a coordinated program: blood vessels widen, immune cells arrive, and chemical messengers amplify the response. This program is essential for survival, yet it is also dangerous. A body that inflames too easily damages its own tissues; a body that inflames too reluctantly invites infection. The circadian system offers a way to keep inflammation on a leash without disarming it entirely. That leash is partly hormonal.

Cortisol is the best-known hormonal brake on inflammation. It is a glucocorticoid produced by the adrenal glands, and it has broad effects: it raises blood glucose availability, helps maintain blood pressure, and suppresses many inflammatory pathways. Clinicians know cortisol because pharmacological versions—prednisone and related drugs—are among the most effective anti-inflammatory treatments ever discovered. What is less widely appreciated is that cortisol is rhythmic. In humans, it tends to rise during the latter part of the night and peak around the time of waking, then decline across the day to low levels in the evening and early sleep. This daily pattern is not just a response to getting out of bed; it is generated by circadian control of the hypothalamic—pituitary—adrenal axis, then shaped by sleep and stress.

The rhythm has a consequence that sounds paradoxical at first: even though cortisol rises toward morning, there is a period in the night when its levels are at their lowest, and during that window inflammatory signals can run relatively less restrained. That does not mean the immune system is "off" during sleep. It means the balance between pro-inflammatory and anti-inflammatory forces shifts with predictable timing. In many inflammatory disorders, symptoms peak when that balance is tipped toward inflammation.

Asthma offers a familiar example. Many people with asthma can tell time by their lungs. Nighttime and pre-dawn hours are notorious for wheeze, cough, and a feeling of tightness. Parents of asthmatic children learn the cruel choreography: a child who seems fine at bedtime can be upright and struggling by early morning. Part of this is mechanical. Airway caliber is not constant; the smooth muscle around the bronchi has its own tone, and that tone varies over the day. Lying down changes lung volumes. Nasal congestion increases at night. Yet immune rhythms are not bystanders. The airways of many asthmatic individuals are primed for inflammation, and inflammatory mediators can fluctuate across the 24-hour cycle, influencing hyperreactivity and mucus production. The result is a disease that does not merely *happen* at night but often *worsens* at a biologically predictable time.

An older medical term, the "dawn phenomenon," is often used for the early-morning rise in blood glucose seen in diabetes, driven by hormonal changes that promote glucose release from the liver. The phrase is sometimes borrowed, informally, for asthma as well: a dawn worsening that can feel like the body is staging a daily ambush. The important commonality is that dawn is not simply dawn. It is a transition point in internal physiology when multiple systems shift phase together: cortisol rises, sympathetic nervous activity changes, airway tone shifts, and immune signals reorganize. If a person's disease process is sensitive to any of these levers, symptoms will appear with a clock-like regularity.

Rheumatoid arthritis (RA) makes the principle even clearer because its hallmark symptom is explicitly temporal: morning stiffness. RA is an autoimmune disease in which the immune system attacks the joints, producing inflammation, pain, and, over time, structural damage. Patients often describe their mornings as negotiations. Fingers feel thick; wrists resist movement; knees seem reluctant to bear weight. Many find that as the day progresses, stiffness eases somewhat, even if pain persists. This is not merely because people "warm up." Inflammatory mediators show their own rhythms, and in RA a key cytokine called interleukin-6 (IL-6) tends to rise during the night and into the early morning. IL-6 is a messenger that can drive feverish feelings, fatigue, and joint inflammation. When IL-6 peaks in the hours before waking, it sets the stage for the particular misery of the first steps out of bed.

A concrete therapeutic insight grew out of these rhythms. Glucocorticoids like prednisone are powerful anti-inflammatories, but long-term use carries risks, including bone loss, metabolic side effects, and susceptibility to infection. Clinicians therefore try to use the lowest effective dose and, when possible, time dosing strategically. In the late 2000s, modified-release formulations of prednisone were developed to release the drug hours after ingestion. The cleverness was simple: take the pill at bedtime so that it releases in the early hours of the morning, around the time IL-6 is surging. Patients in trials experienced reduced morning stiffness and lower early-morning IL-6 levels, illustrating a core idea of chronotherapy: treatment aligned to internal timing can outperform treatment that ignores it, sometimes even at the same total dose.

Chronotherapy is not a new ambition. In the nineteenth century, physicians debated the best times to administer quinine for malaria and digitalis for heart conditions, often relying on clinical impression rather than mechanistic understanding. What is new is that circadian biology provides a rigorous framework for why timing should matter. Drugs are not dropped into a static body; they enter a moving landscape. Absorption in the gut, metabolism in the liver, and the sensitivity of target tissues all vary across the day. For immune-modulating drugs, the landscape is especially dynamic because the immune system itself is rhythmic.

Time-of-day patterns appear not only in chronic inflammatory diseases but also in acute events. One of the most studied is acute myocardial infarction (heart attack). Large datasets have repeatedly shown a morning peak in symptom onset, often rising in the early morning and peaking in the mid-morning hours. This observation has been robust enough to become medical folklore: hearts fail after sunrise. Yet folklore can harden into false certainty. The pattern is real in many populations, but it is not universal, and its meaning is subtle. A morning peak in *occurrence* does not automatically mean a morning peak in *severity*. Contemporary registries sometimes find that while onset is more common in the morning, outcomes such as infarct size or mortality do not necessarily vary strongly by time of day. The clock may influence the probability that a vulnerable plaque ruptures or that clotting becomes more likely, without dictating how much damage ultimately occurs once the event is underway.

What could drive a morning peak in heart attacks? Several rhythmic factors converge. Blood pressure and heart rate tend to rise on waking, increasing mechanical stress on vessel walls. Platelets, the blood fragments that help form clots, show time-of-day changes in reactivity. The balance between clotting and clot breakdown shifts. Cortisol and adrenaline rise, mobilizing energy and tightening vessels. Inflammation, too, contributes to atherosclerosis and plaque stability; it is not hard to imagine that rhythmic immune signals could tip the balance in a vulnerable person. The heart attack data therefore fits the broader story: disease events can cluster at times when multiple physiological levers align in a risk-enhancing direction.

There is an additional twist that reveals how disease can reshape circadian signatures. Diabetes appears to attenuate the classic morning peak in myocardial infarction onset in some analyses. That is not as paradoxical as it sounds. Diabetes alters autonomic function, vascular biology, inflammation, and hormonal rhythms; it can blunt or distort the very signals that create a pronounced daily pattern in healthier individuals. The clock's influence on disease is therefore not one-way. Chronic illness can feed back on circadian organization, flattening rhythms or shifting phases. In that sense, a disease can partially erase the timetable that helped reveal it.

This interplay matters for interpretation. Suppose a clinician reads that asthma is worse at night or that heart attacks peak in the morning. It is tempting to turn those statements into rigid expectations. A patient who wheezes at noon is not "atypical"; they are a person whose triggers, treatment, chronotype, and environment have changed the rhythm. A patient who has a heart attack at midnight has not violated a law of nature. Circadian influences change probabilities, not absolutes. They are more like the tide than the train schedule: they raise and lower the baseline risk, but storms still happen.

The reason symptom timing can be so regular is that the immune system and endocrine system operate as coupled oscillators. A cytokine burst can influence sleep. Sleep loss can increase inflammatory tone. Cortisol can suppress cytokines, but cytokines can also alter cortisol regulation. Even the autonomic nervous system, which adjusts airway tone and blood pressure, is in constant conversation with immune cells in tissues. When these systems are synchronized, symptoms can become more predictable. When synchronization is disrupted—as in shift work, chronic stress, or irregular sleep—

symptoms can become erratic, sometimes worse because the body loses its ability to confine inflammation to safer windows.

One can see the clinical world through this lens and notice small, human-scale consequences. Consider a person with mild asthma who is stable most of the year. They take an inhaled steroid regularly and rarely need a rescue inhaler. Then they take a new job with rotating shifts. Sleep becomes fragmented. Meals shift later. Light exposure at night becomes routine. After a few months, they find that wheeze returns, especially in the early morning after night shifts. It would be easy to blame the workplace air or "getting older." Circadian disruption is a plausible contributor: altered cortisol rhythms, altered immune gating in the airways, and altered autonomic tone can create a background in which the same environmental trigger produces a stronger response. The disease has not changed its identity; the body's timing has changed its threshold.

The deeper implication of timing is ethical as well as biological. If symptoms and risks have rhythms, then "standard" medical practices can accidentally favor certain outcomes. Hospitals commonly administer medications at convenient times for staffing. Blood samples are often taken at dawn. Research studies sometimes collect all samples at one clock time without considering participants' sleep schedules. These practices are understandable; healthcare systems are under pressure. Yet they can blur the signal. A drug trial might miss a benefit because the dosing schedule fights biology. A lab value might be interpreted as abnormal when it reflects normal circadian variation. A patient's complaint of morning stiffness might be minimized as subjective, even though it tracks a measurable cytokine rhythm.

Chronotherapy is not a promise that timing alone can solve disease. It is a reminder that the body is not indifferent to time, and that medicine, for all its sophistication, sometimes behaves as if it were. The most compelling chronotherapeutic successes tend to occur when three conditions are met: the disease has a strong daily symptom rhythm, the drug targets a pathway that is rhythmic, and the dosing can be aligned with that rhythm without increasing side effects. Rheumatoid arthritis with modified-release prednisone comes close to that ideal, and it has the clarity of a clockface: aim treatment at the pre-dawn rise in inflammatory mediators rather than waiting until the patient wakes already stiff and sore.

There is also a quieter application that reaches beyond pharmaceuticals. Patients often discover, by trial and error, that behavior timed to their symptoms helps. People with asthma may adjust evening routines to reduce late-night triggers and ensure controller medications are taken consistently. People with inflammatory pain may plan gentle movement soon after waking to exploit the natural easing of stiffness later in the day. These strategies do not replace medical care, but they respect the reality that the body's vulnerabilities are not evenly distributed across time.

The most unsettling part of symptom timing is how easily it can be ignored. Pain and breathlessness demand attention in the moment; they rarely prompt curiosity about phase relationships between hormones and cytokines. Yet the clock's imprint is often visible to anyone who looks. Morning stiffness, night-time wheeze, mid-morning heart attacks, evening fevers—these are not quirks. They are signatures of a body that allocates defense and restraint in time, sometimes successfully and sometimes not.

A disease, in this view, is not only a faulty molecule or a misguided immune cell. It is a disturbed relationship between a biological process and the timekeeping system that normally contains it. When that relationship is understood, treatment can become less like a blunt intervention and more like a well-timed conversation: quieting inflammation before it crests, supporting airways before they narrow, guarding a vulnerable heart during hours when multiple risks coincide. Time does not cure, but it can be recruited as an ally, and in the small hours of the morning that alliance can make the difference between waking into pain and waking into motion.

Chapter 7

Out of Sync

Modern electricity and global travel have unmoored us from the solar day, turning our internal temporal order into chaos. We investigate the physiological price of this disconnection, from the subtle drag of social jet lag to the profound metabolic confusion caused by working while the world sleeps.

7.1 The Weekend Time Traveler

On Friday night, a quiet act of time travel takes place in millions of bedrooms. The alarm clock, so bossy from Monday to Thursday, is suddenly disarmed. Dinner runs late. A show turns into two. One last scroll becomes thirty minutes. Nothing dramatic happens, and that is part of the trap: the body's clocks do not register the weekend as a holiday. They register it as a sudden shift in the timing of light, food, movement, and sleep.

By Saturday morning, many people have slid into a different time zone without moving an inch. Not by changing how *much* they sleep, but by changing *when* they sleep.

Chronobiologists have a name for the gap between the time your body would like to live on and the time your obligations demand: *social jet lag*. The phrase is deliberately provocative. Jet lag is what you feel after a flight across time zones; social jet lag is what you feel after a commute across calendars. No passport stamps, no airline peanuts, just a workweek that forces one schedule and a weekend that reveals another.

A useful way to see it is to imagine two clocks running in parallel. One is the social clock: school start times, morning meetings, the fact that your neighbor mows the lawn at 9 a.m. The other is biological time: the internal rhythm that decides when melatonin rises in the evening, when core body temperature reaches its minimum, when the brain is most ready for sleep, and when the body expects breakfast. Biological time differs from person to person. Some people feel a pull toward early mornings; others do not begin to feel truly alert until late morning and can work happily into the night. That personal timing tendency is called *chronotype*.

Chronotype is often described with a shorthand that sounds like a personality quiz: larks and owls. The underlying biology is not whimsical. It reflects differences in the timing of circadian phase (when the internal night begins and ends), in how strongly light advances or delays the clock, and in how sleep pressure builds across the day. If you have to wake at 6:00 a.m. for work but your internal night is still in full swing at that hour, you are doing something more strenuous than getting out of bed. You are asking the brain to perform wakefulness at a time it has biologically labeled as "still night."

The simplest way to quantify this mismatch borrows an idea from navigation: the midpoint. Take the time you fall asleep and the time you wake up, and find the halfway point. If you sleep from midnight to 8:00 a.m., your sleep midpoint is 4:00 a.m. It is an imperfect measure, but it captures something surprisingly stable: across many days, people tend to show a characteristic sleep midpoint on days when they are not constrained by obligations. That free-day midpoint is a rough proxy for where their internal night wants to sit.

Now compare the midpoint on workdays with the midpoint on free days. The difference, measured in hours, is the magnitude of social jet lag. It is common to see a difference of one to two hours; for late chronotypes, two to three hours is not rare. A three-hour swing is roughly equivalent to flying from New York to Denver every Friday night and returning every Sunday night, forever, without the scenery.

The weekend feels like recovery because it offers sleep extension. A person who has been waking at 6:00 a.m. all week may sleep until 9:00 a.m. on Saturday, and subjectively that can feel like a rescue. Yet the body does not interpret a later wake time as "paying back" a debt in a simple accounting sense. It also interprets it as a shift in the timing of light exposure (the strongest time cue), a shift in the first meal, a shift in activity, and often a shift in the timing of caffeine and alcohol. The weekend can therefore solve one problem (not enough sleep) while creating another (not enough regularity).

Consider a concrete example. Imagine Maya, a 28-year-old hospital administrator who is not a shift worker. Her job is firmly daytime: she needs to be at her desk by 8:00 a.m. Most nights she turns out the light at 12:30 a.m., but she does not fall asleep quickly. She wakes at 6:30 a.m. on weekdays, hits snooze twice, grabs coffee, and feels human around 10:00 a.m. On Friday night she meets friends for dinner and arrives home at 1:00 a.m. She sleeps until 10:00 a.m. on Saturday. Sunday night she tries to be "responsible," goes to bed at 11:00 p.m., and lies awake, frustrated, until after midnight. Monday morning feels like a betrayal.

This pattern is so familiar that it is often treated as a moral issue: more discipline, fewer screens, better habits. Sometimes those are part of the solution. But for many people like Maya, the problem is not willpower. It is arithmetic. A late chronotype working an early schedule is living with a weekly cycle of misalignment. The weekend is not simply indulgence; it is the only time her sleep is allowed to occur near the phase her internal clock prefers. The trouble is that the "permission" arrives in two-day bursts that repeatedly tug her timing back and forth.

One reason the tug feels so forceful is that the circadian system is not a single metronome. Light arriving in the morning is a strong signal that shifts the central clock earlier (a phase advance) and makes it easier to fall asleep earlier the next night. Light arriving late at night pushes the clock later (a phase delay) and makes it easier to fall asleep later. Weekdays for a late chronotype can contain strong advancing signals: early alarms, bright morning light during the commute, breakfast at a fixed time. Weekends may contain stronger delaying signals: sleeping through the bright part of morning, getting more light late in the evening, eating later, socializing later. The clock is not drifting randomly; it is being trained by two competing weekly regimens.

This is where the metaphor of time zones earns its keep. The body is an anticipatory machine. The liver adjusts enzymes in expectation of meals. The pancreas adjusts insulin secretion patterns. Fat tissue changes sensitivity to hormones that regulate appetite. The cardiovascular system shows daily rhythms in blood pressure and clotting tendency. Many of these rhythms are coordinated, directly or indirectly, by the central clock that is most strongly aligned by light. When behavior repeatedly occurs at times the body labels as "biological night," the rhythms are no longer neatly lined up with the environment.

A weekend schedule shift may sound too small to matter, but biology is not impressed by our intuition about scale. Hormonal systems and metabolic pathways can be sensitive to timing differences of an hour or two, especially if those differences are repeated weekly for years. The point is not that a Saturday sleep-in is toxic. The point is that a *pattern* of weekday restriction plus weekend delay creates a chronic sawtooth in timing that keeps the system from settling.

The term social jet lag was formalized in the mid-2000s by researchers who noticed something that had been hiding in plain sight. When people are allowed to sleep when they choose, their sleep timing shows broad variation. When they must obey work or school schedules, that variation is compressed, and for many people the compression is not gentle. It is a forced shift. If you look at large datasets, late chronotypes often sleep substantially less on workdays and then "catch up" on free days. That catch-up is not mere laziness; it is the body reclaiming sleep that was squeezed out by an incompatible start time. But the catch-up typically comes with a later midpoint, which is exactly the timing shift that constitutes social jet lag.

A historical anecdote makes the social nature of the problem vivid. Before the industrial era, daily life in many places was less governed by a standardized clock. People worked by daylight, seasons mattered more, and sleep could be segmented or flexible, especially outside of urban wage labor. The invention of reliable mechanical clocks did not immediately transform sleep; the transformation came when clocks became instruments of coordination and discipline. Factories needed synchronized starts. Railroads needed standardized timetables. In the nineteenth century, "railway time" helped unify local times into standardized time zones so trains could run safely. That was a triumph of coordination. It also created a world in which the clock on the wall could overrule the clock in the body.

School start times are a modern echo of that transformation. Adolescence is associated with a biological tendency toward later sleep timing, and yet many schools begin early. The mismatch is so pronounced that it has become a public health and educational policy issue in its own right. Social jet lag is not only an adult problem; it is often trained into people early, long before they have any control over their schedule.

Why does this mismatch have metabolic consequences? Partly through sleep duration, but not only. If you cut sleep short, appetite regulation changes: people often feel hungrier, crave more energy-dense foods, and have less self-control around snacks. That is a well-known route from short sleep to weight gain. Social jet lag adds another route: *timing*.

Food is not just fuel; it is information. Eating tells peripheral clocks—especially in organs involved in digestion and metabolism—that it is "daytime" for them. If you eat late at night, you are sending a daytime signal to tissues that may still be on a nighttime program. The body can cope with an occasional late meal; the more relevant question is what happens when late meals and late sleep become a repeating pattern layered onto early alarms and early breakfasts.

Many late chronotypes end up doing something like this: on weekdays, they wake early under duress, drink coffee to push through, eat irregularly or grab quick calories, and then eat a large late dinner because they are finally hungry and finally have time. On weekends, they shift later and eat later still, often with more alcohol. Each element has its own effects, but the timing pattern is the thread that ties them together.

Epidemiological studies have repeatedly found associations between greater social jet lag and higher body mass index, even after accounting for sleep duration. That does not prove causation. People who are socially jet-lagged may differ in other ways: stress levels, commuting, exercise, diet quality, income, and so on. Yet the association is plausible because the mechanisms exist. Experimental work in controlled settings has shown that misalignment between behavioral schedules and internal circadian phase can alter glucose regulation, hormones like leptin, and blood pressure on surprisingly short timescales. In other words, the body does not need decades to notice that its signals are arriving at odd hours; it responds within days.

The lived experience of social jet lag often begins with a feeling that is easy to dismiss: a peculiar dullness on Monday mornings, a sense of being "foggy," and a craving for stimulants. Many people treat this as the normal price of adult life. But Monday is not inherently poisonous. It is simply the day when, for a large fraction of people, the schedule abruptly demands a phase advance. The internal clock cannot jump instantly. The central clock can shift, but it does so gradually, guided by light. When the weekend has delayed someone's timing, Monday becomes a forced attempt to act like an earlier chronotype for at least one day, sometimes two.

An underappreciated detail is that the weekday schedule is not only earlier; it is also more regular. Regularity is usually good for circadian coordination. That is why advice about "keeping a consistent sleep schedule" works for many people. The irony is that the people who most need a consistent schedule are often those who find it hardest to maintain because the consistent schedule on offer is incompatible with their biology. The system then oscillates between two imperfect options: align with biology on free days and suffer socially, or align with society on workdays and suffer biologically.

There is also an emotional cost. Late chronotypes are often labeled as undisciplined, unmotivated, or immature. The stereotype has deep roots; "early to bed and early to rise" is practically a moral slogan. But when a trait is widespread, stable, and partially heritable, moralizing becomes a category error. A society can choose to privilege early timing, but that choice will create predictable losers, and those losers will disproportionately be forced to live in a state of chronic circadian misalignment.

The consequences can be subtle and cumulative. A person with social jet lag may not feel acutely ill. They may simply feel as though their appetite is a little harder to manage, their mood a little more brittle, their workouts a little less effective, their concentration a little less steady. Over years, small shifts in metabolism and behavior can accumulate into weight gain or cardiometabolic risk. Again, none of this means that sleeping in on Saturday is a mistake; it means that the underlying mismatch is not solved by two days of freedom. It is merely expressed.

A revealing way to think about "catch-up sleep" is to split the problem into two parts: *quantity* and *timing*. Catch-up sleep can improve quantity. If someone has accumulated sleep debt, extra hours on the weekend can reduce sleepiness and improve subjective wellbeing. But it can simultaneously worsen timing misalignment by pushing the sleep window later. The best outcome would be to increase quantity without dramatically shifting timing. That is difficult if weekday wake times are very early, because the only way to get more sleep without moving the midpoint is to go to bed earlier—precisely what late chronotypes find hardest to do.

The concept of a sleep midpoint also clarifies why "just go to bed earlier" is often unrealistic advice. If your internal clock is set such that melatonin rises late and your circadian alerting signal is strong in the evening, trying to sleep earlier can feel like trying to nap at noon. You can lie down; you can turn off lights; you can be sincere about your intentions. But sleep is not a behavior you can force directly. It is an emergent state that arises when sleep pressure, circadian phase, and environment align. If those pieces do not line up, effort turns into frustration, and frustration itself becomes arousing.

A small, almost mundane historical note underlines how modern the problem is. Electric lighting, which spread through cities in the late nineteenth and early twentieth centuries, extended the day into the evening and made late hours more usable. That cultural shift disproportionately benefits late chronotypes: it allows more social life, more work, more entertainment after dark. But the morning did not move. In fact, in many places the morning became more rigid as industrial schedules and later school schedules standardized. The result is a squeeze: nights expand, mornings remain fixed, and sleep becomes the variable that gets sacrificed. Social jet lag is, in part, the physiological cost of living in a world where the evening is elastic and the morning is not.

The term "weekend time traveler" is not meant as a scold. It is meant to capture a peculiar feature of modern life: a two-day experiment that repeats every seven days. The body receives one set of time cues for five days and another set for two. The oscillation is not random; it is socially patterned. That is why it clusters: students, service workers, parents with school-aged children, office workers with early meetings. It also explains why some people seem almost immune. If your chronotype is early and your job schedule aligns with it, you may keep similar sleep timing across the week without trying. Your weekend sleep-in might be modest, and your midpoint barely moves. Social jet lag is not equally distributed across the population; it is concentrated among those whose internal timing is late relative to the demands placed on them.

What, then, does it mean to live *against* one's chronotype? It means repeatedly asking the brain and body to perform wakefulness and metabolism at a circadian phase that is optimized for sleep and fasting. It means consuming caffeine when the body expects rising cortisol and alertness, then consuming dinner when the body expects metabolic wind-down. It means treating Monday morning as an isolated hurdle rather than as the predictable consequence of a weekly phase shift. Most of all, it means living in a state where "normal" tiredness becomes indistinguishable from "misaligned" tiredness.

Some people stumble into a solution without naming it: they stabilize timing. Not perfectly, and not by turning into an ascetic. They keep wake time on weekends closer to weekday wake time, even if they allow a modest sleep extension. They seek bright light earlier in the day and dimmer light at night. They avoid making the first meal drift by several hours between weekdays and weekends. These are not glamorous interventions; they are acts of coordination. The goal is not to eliminate pleasure or spontaneity, but to reduce the weekly whiplash that makes the body keep re-solving the same problem.

Others find, painfully, that stabilization is impossible under their constraints. A single parent with two jobs does not "choose" irregularity. A student with late-night study and early classes may have no room to negotiate. A person commuting long distances may have to wake early regardless of chronotype. Social jet lag is therefore not merely an individual health behavior issue; it is an architectural feature of how society schedules time.

A final image is worth holding onto because it prevents the conversation from becoming moral or abstract. Picture two versions of the same person. On Saturday morning, allowed to wake naturally, they open their eyes at a time that makes sense to their brain. They feel oriented. Hunger arrives in an orderly way. The day has a smooth start. On Monday morning, the alarm drags them out of deep sleep, and they begin the day already displaced from themselves. Both versions are real. The difference between them is not character. It is timing.

7.2 Working in the Shadows

At 2:47 a.m. the hospital corridor has a peculiar stillness: the lights are bright enough to flatten shadows, but the building itself feels asleep. A nurse walks quickly past closed doors, pushing a cart that squeaks at every seam in the floor. In one room a monitor chirps; in another, an IV pump complains in a tone that is somehow both urgent and bored. Downstairs, a security guard chews something crunchy. In the break room, a plastic container of pasta sits beside an energy drink. The hands are working, the eyes are open, the world is lit, and yet the body is being asked to do a day's labor during its own night.

Shift work is often described as a lifestyle, a job feature, a necessary sacrifice for a 24-hour society. Physiologically, it is closer to a forced experiment in circadian misalignment. The same biology that makes jet lag miserable after a flight is recruited, repeatedly, without the flight. But shift work is not only a question of sleep. It is also a question of *light* arriving at the wrong time and *food* arriving at the wrong time, and those signals reach into the cardiovascular system, the pancreas, the liver, and fat tissue. A night shift is not just wakefulness in darkness; it is wakefulness in *light* during the biological night, followed by sleep in a world that keeps insisting it is day.

A concrete picture helps. Take someone who rotates onto nights in an intensive care unit: three 12-hour shifts from 7 p.m. to 7 a.m., then a few days off. On the first night, they may still be partially aligned to daytime life. They arrive at work having been awake since morning, and by 4 a.m. their eyelids feel weighted. Their reaction time slows. They snack, because eating offers a brief sense of alertness and comfort. They drink coffee, sometimes late enough that it will haunt their attempt to sleep after the shift ends. After work they drive home in bright morning light, which is exactly the kind of signal that tells the brain, *wake up, it is morning*. They put on sunglasses, but even reduced light still leaks through. They finally collapse into bed at 9 a.m., when their circadian system is sending strong daytime signals to be awake. They sleep lightly, waking to the neighbor's garbage truck or to a delivery notification. They get up at 2 p.m. feeling as if they have slept *near* sleep, rather than actually sleeping.

By the second and third nights, something strange can happen. They may begin to feel more awake during the shift. The body is trying to adapt. But full adaptation is rare, especially for rotating schedules. Days off pull the person back toward a normal social day, with daylight, errands, family dinners, and perhaps children who need attention in the morning. The result is a perpetual in-between state: never fully aligned to night work, never fully aligned to day life. This is where the phrase *working in the shadows* becomes literal. The work takes place under fluorescent ceilings at the time the body expects darkness, and recovery sleep takes place behind curtains at the time the body expects daylight.

One way to appreciate the strain is to remember what the central clock is designed to do. The brain's master pacemaker, located in the suprachiasmatic nucleus (SCN), treats morning light as its primary cue. Morning light says, *anchor the day here*. Night work flips the meaning of light. In many workplaces, the brightest light arrives at midnight, while the commute home delivers outdoor sunlight at the very moment when the worker most needs their internal night to hold steady. For a night-shift worker, darkness becomes scarce, and darkness is not merely the absence of light. Darkness is the signal that permits melatonin to rise.

Melatonin is often marketed as a gentle sleep aid. Biologically it is better understood as a hormone of darkness, a nightly signal that coordinates processes throughout the body. When light hits the retina at night, melatonin secretion is suppressed. The sensitivity is not uniform; it depends on intensity, duration, timing, and wavelength. Short-wavelength (blue-enriched) light is especially potent at suppressing melatonin and shifting circadian phase. Modern workplaces and screens are frequently rich in exactly those wavelengths. A night shift, then, does not only keep you awake. It actively tells the central clock that the night is not night.

The metabolic consequences begin with a simple fact: glucose tolerance follows a daily rhythm. Most people handle sugar better during the biological day than during the biological night. Insulin sensitivity is generally higher earlier in the day, and the pancreas and liver are more prepared for meals then. Eating at 2 a.m. is therefore not metabolically equivalent to eating the same food at 2 p.m., even if the calories are identical. It is like trying to run a factory on its maintenance schedule. The machines will run, but they will not run smoothly.

Shift work pushes meals into the biological night for practical reasons. A person cannot go twelve hours without eating, and many night shifts include a "lunch" break around midnight or 1 a.m. Often the available food is limited to vending machines, cafeteria leftovers, or whatever can be carried from home and eaten quickly. Even when someone brings a carefully planned meal, the timing is still unusual. The liver receives a strong signal: *process nutrients now*. Meanwhile the SCN, entrained by whatever light pattern the person has accumulated, may be broadcasting a different message: *it is night; conserve and repair*. The body is a network of clocks, and night work encourages those clocks to disagree.

Laboratory studies have made this disagreement visible by separating the timing of behaviors from the internal circadian phase. One influential approach uses "forced desynchrony" protocols, in which participants live on an imposed schedule that is too long or too short for the circadian system to follow, so sleep and meals drift across all phases of the internal day. That design allows researchers to compare what happens when a person eats and sleeps during their biological night versus biological day while holding many other factors constant. The results are not subtle: circadian misalignment can reduce leptin (a hormone involved in satiety), raise glucose levels despite higher insulin (suggesting reduced insulin sensitivity), increase blood pressure, and alter cortisol rhythms. These are acute changes, the kind that can be measured after days. They make plausible the long-term epidemiological findings that shift work is associated with higher risk of obesity, type 2 diabetes, and cardiovascular disease.

The cardiovascular system is not a passive plumbing network. It has its own daily programs. Blood pressure tends to dip during sleep in healthy individuals, and the heart rate, vascular tone, and clotting factors vary over the day. When sleep is pushed into the daytime, the expected dip can be blunted. Stress hormones can remain elevated. A night shift is often cognitively demanding, and the combination of fatigue and responsibility raises sympathetic nervous system activity—the "fight or flight" arm that helps you stay alert but also constricts blood vessels and increases heart rate. Add to that the stimulant pattern common in night workers: coffee at 10 p.m., another at 2 a.m., perhaps a third at 5 a.m. Caffeine is not evil; it is a tool. But it is also a drug that can nudge blood pressure and delay sleep, and in a chronically misaligned schedule those nudges can accumulate into a persistent strain.

There is a cultural story here too, and it is not new. Long before the modern hospital and the all-night logistics warehouse, the industrial revolution created night work on a scale that would have been unusual in agrarian societies. Gas lighting and then electric lighting made it possible to run factories after sunset. In the nineteenth century, observers described the unnaturalness of children and adults working through the night in textile mills, sleeping in cramped quarters during the day, and living in a world where daylight became something you encountered mainly on your way to and from work. Reform movements argued about morality, productivity, and exploitation, but the biological argument was not yet available in modern terms. The clocks in the body were still unknown. What was visible, though, was exhaustion and illness. Modern chronobiology has supplied the mechanistic language that those observers lacked.

The most emotionally charged consequences of night work involve cancer risk. The International Agency for Research on Cancer classified night shift work as "probably carcinogenic to humans" (Group 2A), based on limited evidence in humans, sufficient evidence in experimental animals, and strong mechanistic evidence related to circadian disruption. The U.S. National Toxicology Program, in its hazard assessment, concluded high confidence that persistent night shift work that causes circadian disruption increases risk of female breast cancer, while emphasizing that "night shift work" is not a single exposure. It is an overlapping set: light at night, sleep disturbance, altered meal timing, stress, and often other occupational factors. This matters because it prevents an overly neat conclusion. The hazard is real enough to be taken seriously, but the pathway is not a single switch labeled "melatonin" that can be flipped back on with a supplement.

Melatonin sits at the intersection of these concerns because it is both a marker and a messenger of darkness. Suppressing melatonin with light at night is one plausible way that circadian disruption could influence cancer biology, through effects on cell cycle regulation, DNA repair, immune function, and hormonal pathways. Yet focusing only on melatonin risks missing the broader phenomenon: the body's internal timing is a coordination problem. When coordination fails, many systems become noisier. Noisy systems are harder to regulate, and regulation is the difference between a minor perturbation and a chronic disease process.

The worker experiences this not as "internal desynchrony" but as craving, irritability, stomach discomfort, and a kind of cognitive thinning that shows up in small errors: misreading a label, forgetting a step, snapping at a colleague. The gastrointestinal system is especially vocal. Night workers often report indigestion, reflux, constipation, or an unsettled stomach. That is not surprising when you remember that gut motility and digestive secretions follow circadian rhythms. Food arriving at midnight meets a digestive tract whose nightly program may be geared toward slower movement and different enzyme patterns. The gut microbiome also shows daily rhythms, and irregular meal timing can reshape those rhythms. The gut is another clock network, and it protests when its cues become erratic.

A common misconception is that the problem is simply sleep deprivation. Many night workers do sleep less, and that alone carries metabolic risks. But even when sleep duration is protected, the timing mismatch can carry its own physiological costs. That is why someone can sleep eight hours after a night shift and still feel off. The sleep may have occurred at a circadian phase when sleep is lighter and more fragmented. The light exposure during the commute home may have reduced melatonin and delayed the internal clock, while family demands or daytime noise may have truncated the sleep episode. The body may have slept, but it did not necessarily experience a consolidated biological night.

Clinicians have a name for the pattern when it becomes debilitating: shift work disorder. It is characterized by insomnia when trying to sleep and excessive sleepiness when trying to be awake, associated with a work schedule that overlaps the usual time for sleep. Diagnosis often involves careful history-taking and sleep logs, sometimes supplemented by actigraphy (a wrist-worn device that estimates sleep timing). Biological measures such as the timing of melatonin onset in dim light can, in principle, map circadian phase, but in routine practice the lived pattern of sleep and wake tells most of the story.

What can be done? The honest answer is: less than we would like, but more than nothing. Some adaptations are practical rather than heroic. Bright light during the night shift can improve alertness, but it risks pushing the clock later and suppressing melatonin. Conversely, keeping light low during the shift might protect circadian night signals but could increase sleepiness and errors. Workplaces thus face a genuine trade-off between immediate safety and long-term physiology, and the best solution may depend on whether a worker is permanently on nights or rotating.

Permanent night shifts can, for some people, allow partial adaptation if the person also keeps a night-oriented schedule on days off and controls morning light exposure. That is socially costly and often incompatible with family life. Rotating shifts are, biologically, the cruelest design because they prevent adaptation and repeatedly impose phase shifts. If rotation is unavoidable, the direction and speed of rotation matter. Rotations that move in a delaying direction (day $\rightarrow$ evening $\rightarrow$ night) are often easier for the circadian system than those that demand sudden advances, but even "easier" does not mean easy.

Meal timing is another lever. Some night workers find that eating lightly at night and placing the largest meal in the late afternoon or early evening reduces gastrointestinal distress and may support better metabolic control. Yet hunger and workplace culture matter. A shared meal at 1 a.m. can be a social glue in an otherwise lonely shift. Humans are not machines, and metabolic advice that ignores companionship rarely lasts.

There is also a broader social question, one that becomes unavoidable once the biology is understood: who gets to sleep at night? Night work is unevenly distributed. It is common in healthcare, emergency services, transportation, manufacturing, hospitality, and warehousing. These jobs are often essential and frequently underpaid. The physiological burden of circadian disruption thus lands disproportionately on people whose work already carries physical and emotional demands. A society that runs 24/7 does not distribute its nights equally.

The nurse in the corridor at 2:47 a.m. is not merely tired. She is doing skilled work at a biological hour when vigilance is naturally lowest, under lighting that tells her brain it is daytime, eating at times that confuse metabolic clocks, and then trying to sleep while the rest of the world produces noise, light, and obligations. The body can endure a great deal, and many people do this work for years. Endurance, though, is not the same as harmlessness. The cost is often paid quietly: a creeping rise in blood pressure, a waistline that seems to grow despite unchanged habits, an A1c number that inches upward, a sense that health requires constant effort rather than settling into place. In the small hours, under bright lights, the body keeps time the best it can, even when time itself has been rearranged.

7.3 Internal Anarchy

Jet lag has a reputation for being trivial: a grumpy inconvenience cured by strong coffee and a brisk walk in the sunlight. Anyone who has tried to give a coherent presentation after an overnight flight knows the reputation is undeserved. The first day in a new time zone can feel like wearing someone else's body. Hunger arrives at odd hours. Your eyelids droop in the middle of a bright afternoon. At 3 a.m., wide awake, you stare at a hotel ceiling and wonder why the air-conditioning is so loud. By the third or fourth day, you may be sleeping at the "right" local time and still feel strangely off, as if recovery is lagging behind the calendar.

That lingering feeling points to the heart of jet lag: it is not simply sleep loss, and it is not simply the brain failing to catch up. It is the body's clocks falling out of agreement with one another. The word *desynchrony* sounds clinical, but the experience is familiar. When internal signals that normally line up in a predictable order become staggered, physiology turns into a badly coordinated orchestra. The conductor may be adjusting to the new venue, but the strings and brass are still reading yesterday's sheet music.

A concrete example makes this easier to grasp. Imagine Daniel, a consultant who flies from Chicago to Paris for a Monday meeting. The time difference is seven hours. He leaves Saturday evening, sleeps fitfully on the plane, lands Sunday morning local time, and tries to behave like a good traveler: he stays awake, walks outside, and goes to bed at 10 p.m. Paris time. The first night is rough, but by Tuesday he is sleeping more or less normally. From the outside, Daniel looks adapted. Inside, some of his organs may still be living in Chicago.

On Tuesday morning he eats a croissant and drinks coffee at 7:30 a.m. Paris time. His brain is accepting "morning" now; he feels alert enough. Yet his liver and pancreas, which help decide how to handle that breakfast, may still be on a delayed schedule, acting as if it were the middle of the night back home. He experiences this not as a philosophical mismatch but as appetite that does not match his schedule, a stomach that feels unsettled, a peculiar fatigue after meals, or a sense that his energy supply is unreliable.

The reason this happens is that circadian timekeeping is distributed. The suprachiasmatic nucleus (SCN) in the brain is often called the master clock, but the phrase can mislead. The SCN is a master in the sense that it is strongly entrained by light and can coordinate timing across the body through neural signals, hormones, and by shaping behavior (when you sleep, when you move, when you eat). Yet many tissues also contain their own clocks, built from cellular "molecular clockwork"—genes and proteins arranged in feedback loops that generate roughly 24-hour rhythms. These peripheral clocks are not just passive followers. They respond to local cues, especially feeding and fasting, temperature cycles, and activity. The body is better described as a network of clocks than a single timepiece.

This network has advantages. A liver cell does not need to ask the brain every hour whether it should express enzymes involved in fat metabolism; it can run a local schedule. A muscle cell can anticipate daily activity cycles and adjust energy-handling proteins accordingly. But the network architecture also has a vulnerability: when the cues that normally keep the clocks aligned conflict, the clocks can drift apart. Jet lag is one of the cleanest ways to create that conflict, because it abruptly reorders the timing of light, sleep, and meals.

The central clock resets relatively quickly because light is a powerful and direct signal. The retina sends information about light to the SCN, and appropriately timed exposure can shift the clock by advancing it earlier or delaying it later. The speed depends on the direction of travel. Traveling east usually requires a phase advance (going to sleep earlier and waking earlier by local time), which many people find harder. Traveling west usually requires a phase delay (sleeping and waking later), which is often easier. Either way, the SCN can move on the timescale of days.

Peripheral clocks can be slower, and they can be stubborn in a different way. The liver is heavily influenced by feeding times. If you arrive in Paris but continue to eat at odd hours, or if your appetite nudges you toward late-night snacks that align with your home time, your liver receives confusing cues. It may shift partially, then stall, then shift again. Muscle may follow activity patterns. The pancreas may respond to meal timing and sleep timing. Each tissue has its own blend of influences and its own inertia.

This is what is meant by *internal desynchrony*: the internal day is no longer a single coherent phase. Different systems are at different "local times" inside you. You can think of the SCN as rapidly updating its calendar, while other organs are still packing their bags.

A historical anecdote helps underline how modern this problem is.

For most of human history, travel was slow enough that the body could adjust gradually. A merchant moving across Europe by horse or ship would shift light exposure and meal times slowly, allowing the circadian network to drift with the environment. Even early long-distance sea voyages, while exhausting in other ways, rarely forced an abrupt multi-hour time change overnight. The invention that made jet lag possible as a routine human experience was not merely faster travel; it was *time zones* combined with aircraft capable of crossing them quickly. When railroads expanded in the nineteenth century, coordinating timetables across cities became so difficult that standardized time zones emerged as a practical solution. Once time was standardized, a person could suddenly find that "noon" meant something very different a thousand miles away. With jet aircraft in the mid-twentieth century, that difference could be imposed on the body in a single night.

Jet lag is often discussed in terms of the brain: sleepiness, concentration, mood. But the more unsettling aspect is metabolic. When eating occurs at the wrong internal phase, glucose regulation can suffer. Many people notice that their appetite becomes oddly distributed: ravenous at midnight, indifferent at lunchtime. This is not just habit. Hormones involved in hunger and satiety show daily rhythms. So do gut motility and digestive enzyme secretion. When the timing network is scrambled, it becomes harder for the body to predict when energy will arrive, and prediction is the point of circadian organization.

Experimental work has shown just how quickly circadian misalignment can disrupt physiology. In forced-desynchrony studies, in which sleep and meals are deliberately scheduled out of sync with the internal clock, markers of metabolic function shift measurably within days: glucose rises even when insulin is higher, suggesting reduced insulin sensitivity; blood pressure increases; cortisol rhythms can flatten or shift. These are not the slow effects of years of lifestyle. They are acute responses to timing confusion.

At the cellular level, the disruption can look dramatic. When sleep occurs at abnormal circadian phases, the rhythmic expression of many genes in human blood can collapse: the fraction of transcripts showing daily rhythms drops sharply. This matters because circadian regulation is not a decorative flourish. It controls when proteins are built, when repair pathways are active, when immune cells traffic, and when metabolic enzymes ramp up or down. If rhythmic gene expression is blunted, the daily program becomes less distinct, and systems that rely on that program may operate less efficiently.

A key mechanism behind internal desynchrony is the tug-of-war between cues. Light is the dominant cue for the SCN, but food is a dominant cue for many peripheral tissues. Animal studies have shown that feeding schedules can shift the phase of peripheral clocks substantially even when the SCN remains locked to the light-dark cycle. The SCN can insist that it is night, while the liver, receiving food at night, begins to act as if it is day. This uncoupling is not a minor disagreement; it can reach many hours. In effect, the body can be partitioned into sub-time-zones.

That partitioning helps explain a common travel experience: you may be sleeping at the correct local night but still feel that meals "do not land right." Your sleep system has partially adapted, but your metabolic system is still negotiating.

A picture can make the idea concrete. Suppose a traveler experiences a 6-hour eastward time jump. Day 0 is the departure day; Day 1 is the first full day in the new time zone. The SCN begins to re-align quickly under the new morning light. The liver and skeletal muscle, influenced by feeding and activity patterns that may remain irregular, shift more slowly. During the gap, there is a period when the SCN and peripheral tissues are misaligned—a window of internal anarchy in which signals that are normally coordinated become temporally displaced.

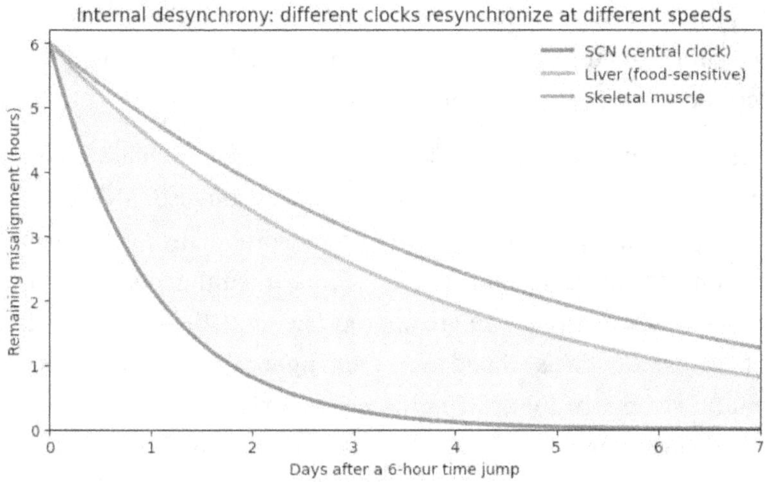

The graph is intentionally schematic, not a claim about any single person's precise re-entrainment rate. Its value lies in the shape: a steep central line and shallower peripheral lines. The shaded region is the uncomfortable territory most travelers recognize, the period when sleep may start to normalize but physiology still feels scrambled.

This internal mismatch also helps explain why "just get sunlight" is good advice and incomplete advice. Morning light can pull the SCN toward the new time, and it can help consolidate sleep at night. But if meals remain mistimed—because work dinners run late, because hunger appears at odd hours, because the traveler grazes on airport snacks—peripheral clocks may lag. Light can adjust one part of the network while food timing drags another.

The reverse is also true. Some travelers attempt to manage jet lag primarily through meal timing, skipping breakfast or delaying food until local daytime in hopes of forcing their body to adapt. Meal timing can indeed provide strong cues to metabolic organs, but it does not replace the central role of light for the SCN. Attempts to control one cue while ignoring the other can leave the network in an even more fractured state, with the central clock on one schedule and peripheral clocks on another.

The subjective symptoms of jet lag often arrive as a bundle: fatigue, insomnia, poor concentration, low mood, gastrointestinal discomfort. It can be tempting to treat them as separate problems: take a sleeping pill for insomnia, antacids for the stomach, stimulants for daytime fatigue. Sometimes those are appropriate, and clinicians who treat jet lag disorder do use timed melatonin, light exposure strategies, and careful sleep scheduling. But the bundle exists for a reason. These symptoms are the experience of a single underlying state: physiological processes that normally take their turns in a coordinated sequence are stepping on one another's lines.

Even the immune system is timed. Many immune functions show daily rhythms, and sleep disruption can alter inflammatory markers. A traveler who feels as if they are "coming down with something" after a long-haul flight is not only reacting to dry cabin air and crowded airports. Their immune timing may be disturbed at the same moment they are exposed to new pathogens. The body is being asked to defend itself while its internal schedule is in flux.

Jet lag also has a curious asymmetry in how it resolves. People often report that they can fall asleep at the correct local night after a few days, yet they still wake too early or feel unrefreshed. That could reflect residual sleep debt, but it can also reflect the fact that the circadian system is not fully aligned. Sleep can be made to occur by exhaustion and by social pressure; that does not guarantee that all internal rhythms are synchronized to the new night.

The point is not that travel is dangerous in itself. The body is adaptable, and for most healthy people jet lag is temporary. The deeper lesson is what jet lag reveals: the body's timing is a network property. When time cues change abruptly, the network does not update all at once. It updates in parts, at different speeds, using different information. The discomfort is the sensation of those parts disagreeing.

Daniel, the consultant in Paris, will likely be fine by the end of the week. He will stop waking at 3 a.m. He will stop feeling hungry in the middle of the night. But if he pays attention, he may notice a quieter shift too: a moment when breakfast suddenly tastes right again, when his stomach feels calm, when the afternoon slump softens. That is not merely him learning the city. It is his internal clocks falling back into step, reassembling a coherent day out of the fragments left by an overnight leap across the map.

Chapter 8

Timing the Cure

Medicine has traditionally focused on the 'what' and 'how much' of dosage, treating the body as a static vessel. We will discover how the timing of a treatment can be the difference between toxicity and cure, revealing why the clock on the wall matters as much as the molecule in the syringe.

8.1 The Morning Hazard Zone

If you had to pick the least forgiving moment for the cardiovascular system, you could do worse than the first hour after waking. Hospitals have long noticed the pattern: more chest pains arriving at dawn, more alarms sounding on cardiac wards in the early morning, more urgent calls that begin with some version of "he was fine when he went to bed." Epidemiology eventually gave that bedside impression a clock face. Across large groups, the onset of heart attacks and strokes is not evenly sprinkled through the day. It rises sharply in the morning, especially in the interval that overlaps with waking, getting out of bed, and beginning the day.

The unsettling part is that nothing obvious has to happen. No marathon. No argument. No dramatic exertion. A person may simply swing their legs out of bed, walk to the bathroom, and in that quiet domestic transition the physiology of night gives way to the physiology of day. The shift is so routine we barely notice it, which is precisely why it matters. The body does not merely *react* to waking; it *prepares* for it. Before the eyes open, networks of clocks have already begun to change the rules of circulation, clotting, and stress signaling. Those changes are usually protective: they ready us to stand up, hunt for food, think quickly, and respond to danger. Yet they also narrow the margin for error in arteries already compromised by atherosclerosis, in hearts already strained by hypertension, or in blood already prone to clotting.

A useful way to picture the morning hazard is as an overlap of three curves. One curve describes the incidence of acute events, such as myocardial infarction (heart attack). A second curve describes cortisol, a steroid hormone that helps mobilize energy and tune the immune system. A third curve describes blood pressure, which typically climbs with waking. The three curves are not identical, but they rise together in the same hours, and that concurrence is not accidental. It reflects a coordinated switch from the night-time body—built for repair, conservation, and horizontal rest—to the daytime body—built for upright action.

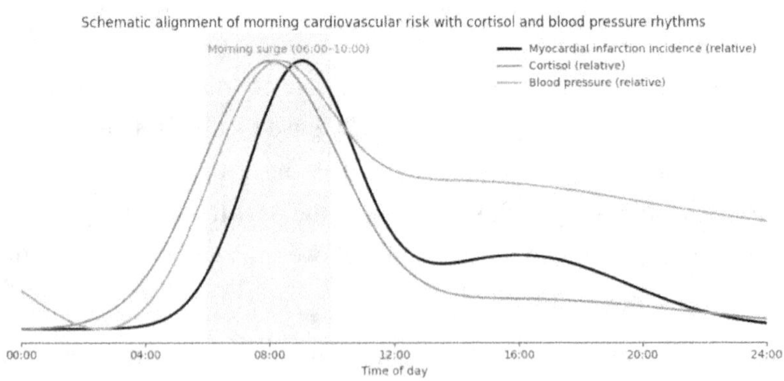

The chart is deliberately schematic. Real people differ: some wake at 5 a.m., some at noon; some have a dramatic blood pressure surge, others barely budge; shift workers can invert or blur these curves. But the principle holds: a coordinated morning transition loads the cardiovascular system in multiple ways at once. The hazard is not one villain. It is a brief convergence of forces that each adds a small amount of strain—until, for a vulnerable plaque or an irritable heart rhythm, the combined strain becomes decisive.

One of the most influential demonstrations of the morning peak came from clinical detective work in the 1980s. In an era when the molecular clockwork inside cells was still largely unknown, cardiologists were already asking a circadian question: *When* do heart attacks begin? Looking carefully at thousands of hospital admissions, they found that symptom onset clustered in the morning, with a clear rise from about 6 a.m. into late morning. That observation was more than a curiosity. It hinted that everyday physiology—not just extraordinary stress—could trigger catastrophe. It also suggested that drugs acting on the nervous system or the clotting system might change not only *how often* events occurred but *when* they occurred.

The timing immediately raised a mechanistic possibility: the sympathetic nervous system. During sleep, the body runs on a quieter setting. Heart rate slows, blood pressure drops, and the balance of autonomic control tilts toward parasympathetic dominance (the "rest and digest" branch). Waking reverses this. Sympathetic tone rises. Adrenal glands release catecholamines. Blood vessels constrict in some regions and relax in others. The heart is told, in effect, to prepare for upright posture and movement.

That surge is not optional. If blood pressure did not rise when you stood up, your brain would briefly be under-supplied with blood and you would faint. The morning increase in pressure is one of the reasons most people can go from lying down to standing and walking without much drama. Yet the same surge increases mechanical stress on arterial walls. If atherosclerotic plaque is present—a fatty, inflammatory buildup with a fibrous cap—a rapid rise in pressure and pulse pressure can increase the likelihood of cap disruption. Once the cap tears, the body treats the exposed material as an injury and rapidly forms a clot. In a coronary artery, that clot can choke off oxygen to the heart muscle: a myocardial infarction. In a cerebral artery, it can cause an ischemic stroke.

So far, that sounds like hydraulics, and in part it is. But the hazard zone is also chemistry.

Cortisol begins to climb before waking, often reaching a peak in the early morning. This rise is not simply a response to opening the eyes; it is anticipatory, shaped by circadian timing and by regular sleep-wake schedules. Cortisol helps mobilize glucose, supports vascular responsiveness to other hormones, and interacts with immune and inflammatory pathways. A body with rising cortisol is a body preparing for activity, and that preparation includes making blood vessels more responsive to constricting signals. In the morning, when sympathetic tone is rising, the vasculature is also becoming more willing to tighten.

Meanwhile, the blood itself shifts subtly toward clot readiness. Platelets—the small cell fragments that help form clots—show daily rhythms in reactivity. Fibrinolysis, the process that dissolves clots, also varies with time of day, and in many people it is relatively less active in the morning than later in the day. The net effect is easy to state in plain terms: in the early morning, the body is often a little more inclined to clot and a little less inclined to undo a clot. Again, that tendency is not inherently wrong. If an injury occurs during daytime activity, rapid clotting helps survival. But when the "injury" is a microscopic tear in a plaque, the same readiness becomes dangerous.

This is where the idea of a distributed timekeeping network earns clinical meaning. Blood pressure rhythms are shaped by clocks in the brain, by hormonal rhythms, by kidney function, by vascular smooth muscle behavior, and by behavior itself—sleep, posture, caffeine, breakfast, commuting. Platelet rhythms involve bone marrow production, platelet aging in circulation, and signaling pathways in platelets that can be entrained indirectly by hormones and activity. The morning hazard zone is not controlled from one central dial. It is an emergent property of many oscillators acting in concert.

A concrete example helps. Imagine two mornings.

In the first, a healthy young adult wakes naturally. Overnight blood pressure has dipped. On waking, blood pressure rises, but arteries are flexible, plaques are absent, and the endothelial lining of vessels (the thin, reactive layer of cells that controls vessel tone and inflammation) is intact. Platelets are ready to do their job but not primed by chronic inflammation. The morning transition is brisk but safe.

In the second morning, an older adult with untreated or undertreated hypertension wakes abruptly to an alarm. Overnight dipping may be blunted, and morning pressure may surge higher. The coronary arteries carry plaques with thin caps. The endothelium is less able to release protective nitric oxide. There may be heightened baseline inflammation from diabetes, smoking, or chronic stress. The same normal morning program runs, but it runs on a riskier substrate, like a well-rehearsed dance on a cracked floor.

The timing of medication intersects this physiology in an almost theatrical way. Consider aspirin, a drug so ordinary it sits in kitchen drawers. Low-dose aspirin irreversibly inhibits platelet cyclooxygenase-1, reducing thromboxane A_2 production and thereby diminishing platelet aggregation. In large randomized data sets, aspirin reduced the incidence of myocardial infarction, and intriguingly it seemed to flatten the morning peak more than other parts of the day. The implication is not that aspirin "knows" the time; rather, platelet-driven mechanisms contribute disproportionately to the morning hazard. When platelet function is damped, the hours that were previously most platelet-dependent lose some of their special danger.

Beta-blockers tell a parallel story. By blocking adrenergic receptors, they blunt the cardiovascular effects of sympathetic activation. In observational analyses of the morning peak, the circadian pattern appeared less evident among patients already taking beta-blockers. Again, the medication does not erase the circadian program of waking, but it reduces one of its sharpest edges: the adrenergic surge.

These examples show why timing matters even when the drug is taken once daily. If a medication's concentration is highest when the hazard is highest, it may have its greatest protective impact. If it peaks at a safer hour and fades during the morning surge, protection may be thinner when needed most. That simple logic underlies much of chronotherapy: not mystical alignment with the cosmos, but a practical attempt to match pharmacology to predictable biology.

Yet the story is more complicated than "take everything at night." The morning hazard zone is real, but bodies differ in their internal time. The most important clock cue is light, and light patterns differ between people and seasons. Behavior also matters: someone who wakes, stays in bed, and eases into the day may experience a gentler hemodynamic transition than someone who jolts up and rushes into cold air. Sleep disorders alter the terrain. Obstructive sleep apnea repeatedly activates the sympathetic nervous system during the night and can abolish the normal blood pressure dip. For such a person, the problem is not merely the morning surge; it is a whole night of physiological "micro-awakenings" that keep the cardiovascular system on alert. In that context, morning may still be dangerous, but the reason is not a clean circadian crescendo; it is a night of accumulated strain.

Even posture has a circadian plot twist. Humans are horizontal for hours, and then suddenly vertical. The act of standing shifts blood volume downward, requiring rapid adjustments to maintain cerebral perfusion. The baroreflex—pressure sensors in the carotid sinus and aortic arch—responds by increasing heart rate and vascular tone. Those reflexes are themselves influenced by circadian state and by hormones. A person whose reflexes are dulled by age, autonomic neuropathy (as can occur in diabetes), dehydration, or certain medications may wobble between too much and too little pressure during the morning transition. Ironically, the same systems that protect against fainting can overshoot and strain the heart.

The morning hazard also exposes a subtle conceptual point: "circadian" and "after waking" are intertwined but not identical. Some rhythms track the internal clock even in constant conditions, while others are tightly linked to sleep, posture, and activity. The morning peak in events likely reflects both. There is a circadian component: physiological readiness for daytime begins before waking, and some variables shift even if behavior is held constant. There is also a wake-related component: standing up, starting to move, experiencing stress, and encountering environmental stimuli are event triggers. In real life, they occur together, which is why disentangling them is difficult and why the hazard feels so consistent.

One historical anecdote from cardiology makes the practical stakes vivid. In the decades when physicians first began to recognize circadian patterns in cardiovascular events, the default model of medicine was timeless: a drug had a dose, a half-life, and an effect, and the clock on the wall was mostly a matter of convenience. But clinicians began noticing that some treatments seemed to work best when they covered the morning hours. Long-acting formulations of antihypertensives were partly motivated by the desire to avoid gaps in blood pressure control during early morning. The shift was not philosophical; it was forced by admissions. When the busiest time for myocardial infarction onset is the same time the body naturally increases blood pressure and platelet activity, a 24-hour drug that quietly wears off at 5 a.m. becomes more than a pharmacokinetic quirk. It becomes a design flaw.

This is also why ambulatory blood pressure monitoring—measuring blood pressure across day and night—changed hypertension care. Clinic readings taken at 2 p.m. can miss the nocturnal dip status and the size of the morning surge. Two patients may have the same office blood pressure yet very different 24-hour patterns: one with a healthy dip and a modest rise, another with a high night-time pressure and a sharp morning surge. Those patterns carry different risks and may call for different strategies. Timing is not an aesthetic preference; it is part of diagnosis.

The morning hazard zone, then, is not a single phenomenon but a synchronized handoff. The endocrine system raises cortisol and other hormones to mobilize energy. The autonomic nervous system shifts toward sympathetic activity to support posture and readiness. The vasculature responds with increased tone and reactivity. The blood tilts toward clot formation and away from clot breakdown. Add to this the prosaic realities of waking—the first cigarette, the first coffee, the rush, the cold air, the climb up stairs—and the dawn hours become a narrow bridge. Most people cross it daily without incident, which can make the danger seem abstract. But for a susceptible artery, it is the moment when many small pushes arrive at once.

Thinking in terms of distributed timekeepers changes the emotional tone of prevention. It suggests that "risk" is not only a static property of a person (age, cholesterol, genetics) but also a dynamic property of *when* that person meets their own physiology. A plaque does not rupture at random. It ruptures at a time when mechanical stress, vascular responsiveness, inflammatory state, and clotting propensity align unfavorably. The morning is one of the few windows where such alignment is reliably common.

The practical corollary is both hopeful and sobering. Hopeful, because a window implies leverage: if risk concentrates in a few hours, then well-chosen interventions that cover those hours could prevent disproportionate harm. Sobering, because the same concentration means that a missed dose, a poorly timed formulation, or a disrupted sleep schedule may have outsized consequences at precisely the time when the body is least forgiving. The morning hazard zone is the ordinary day beginning—and, in a vulnerable cardiovascular system, it is the day's most consequential experiment in timing.

8.2 The Vulnerability Window

Chemotherapy has always carried an uncomfortable double meaning. The word promises a chemical weapon against cancer, yet every patient knows that the weapon does not recognize moral boundaries. It injures hair follicles, gut lining, bone marrow, nerves. It can be lifesaving, and it can be brutal. That brutality is not an accident or a failure of compassion; it is a direct consequence of what cancer is. Most classic chemotherapies are not "anti-cancer" in the narrow sense. They are anti-division, anti-DNA, anti-metabolism—aimed at the very processes that make cells multiply. Tumours multiply relentlessly, but so do some of the body's healthiest tissues.

The practical problem is often described with a crisp phrase: the therapeutic index, meaning the gap between a dose that harms the disease and a dose that harms the patient. In infectious disease, antibiotics can have a generous therapeutic index because bacteria are different from us. In oncology, the difference is more subtle. Many cancers are, in molecular terms, warped versions of our own tissues. The therapeutic index can be narrow enough to feel like a high-wire act:

dose too low and the tumour survives; dose too high and the patient cannot tolerate the treatment.

Timing offers a different kind of leverage. Instead of changing the drug, it asks whether we can change *who* is vulnerable *when*. Healthy tissues do not divide at random. They follow schedules shaped by the daily cycle of feeding and fasting, activity and rest, hormones and local clocks in each organ. Cancer cells, by contrast, often lose the discipline of coordinated timing. They may still have internal rhythms—some tumours retain functional clock genes, others distort them—but they are less likely to respect the body's communal timetable. That mismatch creates a "vulnerability window": periods when healthy tissues are relatively protected, while tumour cells remain exposed.

The idea sounds almost too neat, like a trick in a puzzle book. Yet it comes from the stubborn, material fact that a cell's sensitivity to chemotherapy depends on what it is doing. A drug that blocks DNA synthesis will do the most damage when a cell is copying its DNA. A drug that disrupts mitosis (the division process) will be most lethal when a cell is attempting to split. If a healthy tissue tends to schedule much of its division for a particular part of the day, then giving the drug at the opposite part of the day can reduce harm without changing the amount of drug.

For anyone who has watched chemotherapy administered, the obstacle is obvious: hospitals and clinics run on human timetables. Infusions are scheduled when staff are present, when chairs are available, when pharmacy can prepare the dose. "Time of day" is treated as logistics. Chronotherapy treats it as biology.

A concrete image helps before the details: imagine a gate that swings open and shut over each 24-hour cycle. When the gate is open, a healthy tissue is in a vulnerable state—many cells are cycling, DNA is being replicated, and a cytotoxic drug will hit hard. When the gate is shut, the tissue is comparatively quiescent and more resistant. Now imagine tumour cells as people who do not wait for the gate. They move through the space at all hours. If you can time the drug so that it strikes when the healthy gate is shut, you spare the healthy tissue while still catching a fraction of the tumour population. The tumour does not have to be fully exposed; you only need to widen the therapeutic index enough that treatment can be delivered at effective intensity.

A schematic **'gate' model of chronotolerance:** healthy tissues cycle in a coordinated window; tumours are less synchronized.

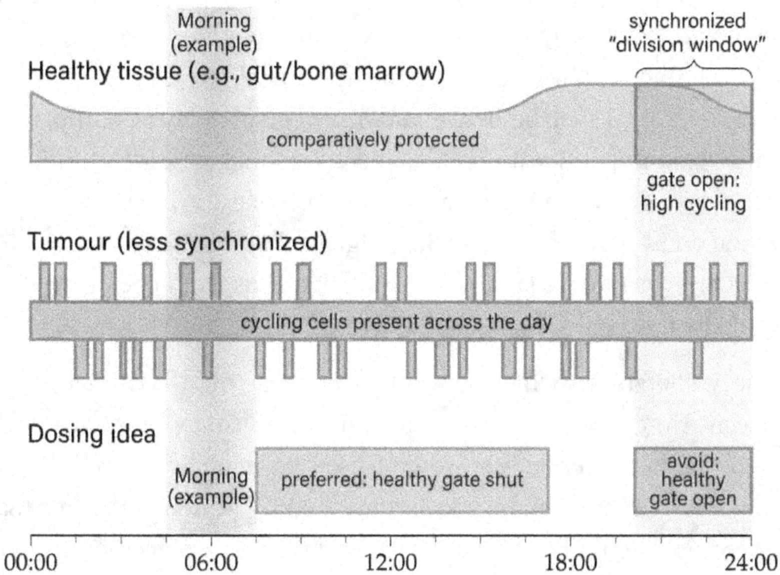

This diagram is conceptual: real tissues differ in phase, amplitude, and patient-to-patient timing. The aim is to widen the therapeutic index by exploiting predictable windows of relative healthy-tissue tolerance.

The gate metaphor is not just a cartoon. It matches what is known about some of the most vulnerable tissues in chemotherapy: the lining of the gastrointestinal tract and the bone marrow. The gut epithelium renews itself rapidly. Stem cells in the crypts of the intestine divide, their progeny migrate, differentiate, and are shed. Bone marrow continuously produces blood cells, and the precursor cells that do this work are among the most actively dividing in the body. Damage them and neutrophils (a kind of white blood cell) can plummet, leaving a person vulnerable to infection. Damage the gut lining and the result is mucositis, diarrhoea, pain, dehydration. These are not mere inconveniences; they can force dose reductions or delays that compromise cancer control.

Why should these tissues have daily timing at all? One reason is brutally practical: digestion and feeding are not evenly distributed. Even in modern life, many people eat most of their calories during daylight. The gut faces recurring waves of mechanical stress, changing pH, bile, microbial metabolites, and dietary components. Coordinating repair and renewal with predictable cycles makes sense. Similarly, bone marrow output is connected to hormones, sympathetic nerve input, and systemic inflammatory signals that themselves have daily structure. A tissue that renews rapidly benefits from deciding when to do its most hazardous internal work.

The vulnerability window therefore depends on a relationship between two clocks: the cell cycle and the circadian cycle. The cell cycle is the series of steps a cell takes to duplicate its DNA and divide. The circadian cycle is the roughly 24-hour program that coordinates physiology across the body. In many healthy tissues, these cycles talk to each other. Circadian regulators can modulate the expression of proteins that drive or brake the cell cycle, and the cell cycle can feed back on metabolic rhythms. The result is not that every cell divides at precisely 3:17 a.m., but that populations of cells have preferred times for entering sensitive phases.

Cancer scrambles that. A tumour is often described as a disease of uncontrolled proliferation, but the deeper problem is uncoordinated proliferation. Mutations in oncogenes and tumour suppressors not only push cells to divide; they can decouple division from the usual environmental checks. A tumour may also develop chaotic blood supply and regions of hypoxia (low oxygen), altering metabolic timing in irregular patches. Some cancers may retain a circadian rhythm, but it may be weak, shifted, or out of sync with the host. For chronotherapy, that disorganization can be exploited. If the tumour does not share the healthy tissue's protected interval, then there is a chance to strike when the patient can best tolerate the strike.

The romantic version of the story is that one could simply schedule chemotherapy at the "right" hour and cure cancer with fewer side effects. The reality is less tidy, and that is part of what makes the field honest. Chronotherapy is a strategy for *tilting* odds, not abolishing trade-offs. The same drug can have multiple toxicities, and those toxicities may peak at different times because different tissues have different clocks. A schedule that spares the gut may stress the bone marrow, or vice versa. And tumours are not one thing; they are ecosystems of subclones with different growth rates and different vulnerabilities. Timing can be one handle among several, but it is not a magic handle.

A historical example shows both the promise and the complexity. In the 1990s, the oncologist Francis Lévi and colleagues pursued chronomodulated chemotherapy for metastatic colorectal cancer using combinations such as oxaliplatin and 5-fluorouracil (5-FU) with leucovorin. The key point was not merely *when* a bolus was given, but that the infusion rate itself could be shaped across 24 hours, like turning a faucet up and down according to a programmed curve. This required technology that looks mundane now but was conceptually daring then: portable pumps that could deliver drugs in a circadian pattern while patients went about their day.

Why bother with such complexity? Because the toxicity-limiting tissues were believed to have rhythms that could be respected. If the same total daily dose could be delivered with fewer side effects by shifting the peaks of delivery into safer hours, the regimen might be intensified or sustained longer. In early randomized work, chronomodulated regimens showed improved tolerability in some settings and made a compelling case that the clock mattered at the bedside, not only in laboratory animals.

Later, larger trials revealed the hard part: the benefits were not universal. In a major phase III trial led by Sylvie Giacchetti and Georg Bjarnason (2006), chronomodulated delivery altered the *pattern* of toxicity—diarrhoea was more prominent in one schedule, neutropenia in another—without delivering a clear survival advantage across the full cohort. The result was not a refutation of timing, but a warning against treating it as a one-size-fits-all improvement. If chronotherapy shifts which tissue takes the hit, it can still be clinically valuable, but it demands careful matching of schedule, drugs, and patient biology.

The most interesting lesson from such trials is that timing can change the *dose-limiting factor*. In oncology, the "maximum tolerated dose" is often constrained by whichever organ system fails first. If a chronomodulated schedule protects one tissue, clinicians can sometimes raise the dose until another tissue becomes the limiting problem. That means timing does not simply reduce toxicity; it can redistribute it. The therapeutic index is not just widened; it is reshaped.

There is also a subtler dimension that becomes visible only when you stop thinking of drugs as static quantities in a bag. Even if the infusion rate is constant, the body's handling of a drug can vary over the day. A striking example comes from studies measuring plasma levels of 5-FU during constant-rate infusion, where concentrations were nonetheless higher at some times and lower at others. The patient receives the same drip, but the internal exposure is not flat.

Clearance varies with liver blood flow, enzyme activity, transporter expression, and other rhythmic processes. In other words, the clock can create a vulnerability window even when clinicians think they have eliminated timing by using a steady infusion.

This helps explain why patients sometimes report that side effects arrive with a daily pattern during multi-day infusions: nausea that worsens at a particular hour, fatigue that crests in the afternoon, mouth soreness that feels like it "turns on" at night. Some of that pattern is behavioural and contextual, but some is pharmacology interacting with rhythmic physiology.

Chronotherapy, at its best, treats this as an engineering problem grounded in physiology. You identify the most dangerous toxicity, the tissue responsible, and the time window when that tissue is most sensitive. You then tune the delivery schedule to avoid that window while still achieving sufficient exposure to the tumour. The schedule might involve shifting a peak to the late afternoon, lowering the rate overnight, or delivering a higher fraction at times when detoxification pathways are more active. It is less like choosing a calendar appointment and more like shaping a waveform.

Yet a patient is not a waveform. The more carefully one looks, the more "time of day" dissolves into personal time. Two people may both receive an infusion that peaks at 4 a.m. is, but for one of them 4 a.m. is a deep biological night, while for another it is effectively late evening due to insomnia, shift work, or delayed sleep phase. Chronotype—the tendency to feel alert earlier or later—is not just a personality trait. It reflects differences in the phase of internal clocks relative to social time. When treatment is timed to the clock on the wall rather than the clock in the body, the intended vulnerability window may miss its target.

This is where the idea of distributed timekeepers becomes clinically consequential in a new way. The tumour has its timing, the gut has its timing, the bone marrow has its timing, the liver has its timing. They do not all move in perfect synchrony, and disease can desynchronize them further. A person with cancer may have disrupted sleep, altered feeding, inflammation, and medications such as steroids that shift rhythms. Even the tumour itself can change the host's systemic rhythms through cytokines and metabolic demands. The neat cartoon gate becomes a moving gate, and sometimes a warped gate.

So the practical question becomes less "What is the best time for chemotherapy?" and more "How do we estimate *this patient's* vulnerability window for *this regimen*?" That estimation can draw on several clues. Sleep-wake timing is a coarse proxy. Core body temperature rhythms and melatonin onset are more direct markers but less convenient. Continuous actigraphy can approximate rest-activity cycles. Some labs are exploring gene-expression rhythms in blood as a readout of internal phase. None of these is yet routine in oncology clinics, but the trajectory is clear: if timing can meaningfully widen the therapeutic index, then measuring internal time becomes part of precision medicine.

Even without sophisticated biomarkers, timing already appears in everyday clinical improvisation. Oncologists learn which side effects are dose-limiting in a given patient and adjust schedules accordingly: shifting anti-nausea medications, changing steroid timing to preserve sleep, choosing infusion times that reduce clinic-day fatigue. These are often framed as comfort measures, but they are also a kind of informal chronomedicine. They acknowledge that tolerance varies with time, that sleep matters for resilience, that appetite and hydration have rhythms, that the same person is not physiologically identical at 9 a.m. and 9 p.m.

The vulnerability window is therefore both a biological fact and a clinical attitude. The fact is that cycling healthy cells can be protected by timing, and tumours can sometimes be hit more cleanly because they are less coordinated. The attitude is a willingness to treat time as a dosing parameter—not a footnote—and to treat the patient's day as part of the therapeutic landscape.

One last image brings the idea back to human scale. A patient with colorectal cancer sits with a portable pump in a small shoulder bag, tubing disappearing under clothing, the drug entering the bloodstream not as a single dramatic moment but as a quiet pattern repeating through night and day. To the patient, the pump may feel like a tether. To the body, it is a timed signal arriving into a network of clocks. The tumour is not politely waiting for office hours, and neither are the gut's stem cells or the marrow's precursors. The vulnerability window is the slim interval when an ancient daily program of tissue renewal can be turned, briefly, into protection.

8.3 Pharmacology's Fourth Dimension

A drug label looks reassuringly timeless. "Take once daily." "Take with food." "Avoid grapefruit." The implication is that the pill is a stable object with a stable effect, and that the only thing that changes is the patient's willingness to swallow it. Yet anyone who has taken the same medication at different hours knows the experience can shift: a sedative that feels too strong when taken late at night, a stimulant that lingers uncomfortably into evening, a blood-pressure tablet that leaves a person lightheaded on a rushed morning but feels gentle on a quiet weekend.

Pharmacology has long described three dimensions: the drug, the dose, and the body. Time adds a fourth. Not merely the time *since* the dose (the familiar rise and fall of concentration), but the time *of day* when that dose lands in a body whose absorption, distribution, metabolism, and excretion are not constant. The heart attack that clusters after waking and the chemotherapy toxicity that changes with infusion timing are both reminders that physiology is scheduled. The same is true for the machinery that handles drugs, and much of that machinery lives in the liver.

The liver is often portrayed as a chemical factory with one job: detoxify. That is true in spirit but misleading in tone. Detoxification makes it sound like the liver politely mops up poisons. In reality, the liver is a switching yard where molecules are tagged, dismantled, rebuilt, and rerouted. Some drugs are inactivated there, some are activated (prodrugs turned into working forms), and some are transformed into metabolites that have their own effects and side effects. The liver also decides, indirectly, how much of an oral dose ever reaches the rest of the body. This is the phenomenon of *first-pass metabolism*: blood from the gut flows straight to the liver, so an oral drug can be substantially processed before it has a chance to circulate widely.

First-pass metabolism is one of those ideas that seems obvious once you know it, yet it took the development of modern pharmacokinetics to make it intuitive. A historical irony is that early drug discovery often relied on effects that were easy to observe but hard to quantify, and the liver's role could remain invisible. As analytical chemistry improved, particularly in the mid-twentieth century, researchers began measuring drug levels in blood and noticing that route mattered in ways that could not be explained by dose alone. The same amount of a compound could be potent when injected and surprisingly weak when swallowed. The body was not being capricious; it was applying a filter, and the filter had its own rules.

Those rules are not fixed across the day. The liver receives rhythmic signals from the central circadian system and from behaviour: feeding patterns, sleep, activity, body temperature, hormones. Liver cells (hepatocytes) also contain their own molecular clocks, and those clocks regulate gene expression in a daily pattern. The consequences can be practical. At one hour, the liver may express higher levels of a drug-metabolizing enzyme; at another, it may express more of a transporter that pumps drugs into bile; at another, it may simply have different blood flow or different redox state, subtly shifting how rapidly a drug is processed.

This rhythmic handling of drugs is often grouped under the term *circadian pharmacokinetics.* Pharmacokinetics is the study of what the body does to a drug: how it absorbs, distributes, metabolizes, and eliminates it. Circadian pharmacokinetics adds the claim that these processes can oscillate over roughly 24 hours. The claim is not mystical; it is a straightforward extension of circadian biology into clinical chemistry. If enzymes and transporters rise and fall, then clearance rises and falls. If clearance rises and falls, then drug exposure rises and falls. And exposure is what tissues feel.

A useful way to keep the logic clear is to separate two ideas that are easily muddled.

One is *chronoefficacy*: a drug works better at one time of day than another.

The other is *chronotoxicity*: a drug causes more side effects at one time of day than another.

Both can occur even if the dose is identical, and both can occur for two broad reasons. The first is that the *target* changes its sensitivity over the day (a receptor, an ion channel, an immune pathway). The second is that the *drug level* changes over the day because absorption or clearance changes. The liver is central to the second reason, and it sometimes quietly contributes to the first as well, because metabolism and signaling are tightly entangled.

Much of this comes down to the liver's core toolkit: the cytochrome P450 enzymes, a large family often abbreviated CYP. These enzymes perform oxidation reactions that make many molecules more water-soluble and easier to eliminate. Different CYP enzymes specialize in different chemical motifs; some are famously important for common medications. The point for timing is that CYP expression and activity can be rhythmic. Molecular circadian regulators control transcription through recognizable DNA elements, and the liver's detox genes sit downstream of that clock-controlled circuitry. The details can be technical—E-boxes, D-boxes, nuclear receptors such as REV-ERB and ROR that act as transcriptional brakes and accelerators—but the everyday translation is simple: the liver does not run the same settings at 3 a.m. and 3 p.m.

Transporters add a second layer. Enzymes transform drugs; transporters move them. In the gut, transporters can influence absorption. In the liver, they influence whether a drug is taken up from blood, pumped back out, or secreted into bile. A well-known example is P-glycoprotein, which can pump a range of drugs across membranes and thereby alter their effective concentration in tissues. Transporters, too, can be rhythmically expressed. Even without changing the chemistry of a drug, the body can change the *routing*.

If this sounds like too many moving parts, consider a concrete case where timing became part of ordinary prescribing: statins.

Statins reduce cholesterol synthesis by inhibiting HMG-CoA reductase, a key enzyme in the pathway. The liver synthesizes cholesterol in a daily rhythm, and in many people it peaks during the night. That matters most for statins with shorter half-lives: if the drug is taken in the evening, peak drug levels overlap more closely with peak synthesis. This does not mean morning dosing is useless, and it does not mean every statin must be taken at night. Longer-acting statins provide more even coverage, and clinical practice often prioritizes adherence over fine-tuning. Still, the statin story illustrates a broader principle: a drug can be more rationally timed when its target process has a predictable circadian pattern.

The statin example is also emotionally revealing in a quiet way. It shows how chronotherapy can feel less like an exotic subfield and more like common sense. If the body performs a particular biochemical job at night, then doing the chemical interference at night is simply meeting the body where it is. The fourth dimension is not an added complication for its own sake; it is an invitation to stop pretending the body is the same at all hours.

A second, more sobering example comes from cancer pharmacology, where timing is not merely a matter of optimizing benefit but of avoiding harm. Even when an anticancer drug is infused continuously, internal exposure can still oscillate. Plasma levels of 5-fluorouracil (5-FU), for instance, have been observed to show time-of-day variation during constant-rate infusion, with higher concentrations in the early morning and lower concentrations later in the day. That observation is a small shock to the intuition of clinicians and patients alike. A steady drip feels like it should produce a steady internal state. The body disagrees. Rhythms in metabolism and clearance can imprint a daily wave onto what appears, externally, to be a flat line.

Once you accept that, many clinical puzzles look different. A patient's nausea that predictably worsens at a particular hour during multi-day therapy is no longer dismissed as "anticipation" or mere coincidence. It may reflect a genuine peak in exposure or sensitivity. Likewise, dose-limiting toxicity may emerge not only from cumulative burden but from recurring daily peaks that repeatedly strike a vulnerable tissue at its most sensitive hour.

The implications extend far beyond oncology. Consider antihypertensives, medications taken by hundreds of millions of people worldwide. Blood pressure has a circadian pattern: it often dips during sleep and surges after waking. That makes timing seductive. If the hazard of cardiovascular events rises in the morning, then perhaps taking blood-pressure medication at bedtime could blunt the morning surge and reduce risk.

This idea has been a source of real controversy in recent years, which is itself revealing. It shows how difficult it is to translate circadian logic into population-level guidance.

Some studies and trials have reported dramatic benefits of bedtime dosing, suggesting that moving antihypertensives to the evening could substantially reduce cardiovascular events. One large body of work, often associated with the Hygia Chronotherapy Trial, attracted intense attention partly because the reported effect sizes were so large that they seemed to promise an unusually simple improvement: the same pills, different hour, far fewer heart attacks and strokes. That promise met skepticism, and questions were raised about study conduct and reporting, including formal expressions of concern. Meanwhile, large pragmatic randomized trials such as TIME (which compared evening versus morning dosing in a real-world setting) did not find major differences in cardiovascular outcomes. More recent pragmatic work, including the BedMed studies, has similarly emphasized that the expected large advantage of bedtime dosing may not appear in broad populations, at least not in a way that cleanly translates into a universal rule.

The debate can look, from the outside, like scientists arguing over trivialities. It is the opposite. It is the fourth dimension colliding with medical culture.

A universal bedtime-dosing recommendation would require several things to be simultaneously true. The morning surge would need to be a dominant driver of events across patients. The drug's pharmacokinetics would need to be such that bedtime dosing reliably increases protective coverage in the morning relative to morning dosing. Side effects such as nocturnal hypotension (dangerously low blood pressure during sleep) would need to be acceptably rare. Most importantly, "bedtime" would need to mean something biologically consistent across individuals. It does not. Bedtime differs by chronotype, work schedule, insomnia, caregiving, and illness. Even the pattern of blood pressure dipping differs: some people are "dippers," some "non-dippers," and some show higher blood pressure at night.

So the question becomes less "Is bedtime dosing always better?" and more "Which patients, with which drugs, under which circadian and behavioural conditions, might benefit?" The fourth dimension rarely rewards one-size-fits-all answers. It pushes medicine toward stratification: identifying who has a large morning surge, who is prone to night-time hypotension, who has sleep apnea, who has shifted schedules, who takes a medication with a short half-life that truly changes morning coverage when moved.

If the complexity feels disappointing, it helps to remember what is being asked. The standard model of prescribing already accepts enormous simplifications: that a clinic blood pressure reflects a person's usual blood pressure; that a single daily dose fits all; that adherence will be similar regardless of timing; that side effects are stable across the day. Circadian pharmacology challenges those simplifications, which means it also inherits all the messy variability that the simplifications were hiding.

This brings us back to the liver, because it is a major source of that variability. The liver does not simply tick with the clock; it responds to food timing with particular sensitivity. A late dinner, a skipped breakfast, nocturnal snacking, intermittent fasting, shift work—these can all shift liver rhythms relative to the light-dark cycle. In animal studies, feeding schedules can strongly entrain the liver even when the central clock remains aligned with light. In people, the patterns are more complex, but the principle matters: two individuals can share the same sunrise and yet run different hepatic schedules because their meals occur at different times.

That has consequences for drug exposure. A medication taken "in the morning" might be taken into a liver that is still in a metabolic mode closer to night if a person's feeding and sleep are delayed. Or a medication taken "at night" by someone who works overnight might coincide with their biological daytime. The clock on the wall is a poor proxy for the clock in the hepatocyte.

There is also the hospital problem. Modern medicine is proud of being 24/7, yet it is not evenly 24/7. Wards have rhythms: medication rounds, blood draws, meal deliveries, shift changes, noise peaks, light exposure at odd hours. A drug that is supposed to be timed carefully may be administered when staffing allows. A patient who would normally sleep in darkness may sleep under fluorescent interruptions. The liver's rhythms respond to these cues. A hospital can therefore unintentionally create circadian misalignment that changes drug handling, exactly in the patients for whom drug safety margins are narrowest.

All of this can sound as though chronopharmacology is a counsel of despair: endless individual differences, shifting schedules, a liver that refuses to behave like a stable machine. The more generous interpretation is that time-of-day effects are not rare quirks. They are a predictable consequence of living in a body designed for day and night.

A simple schematic can anchor the idea without drowning in details. Enzyme expression rises and falls; clearance rises and falls; the same dose yields different concentration curves depending on when it is taken.

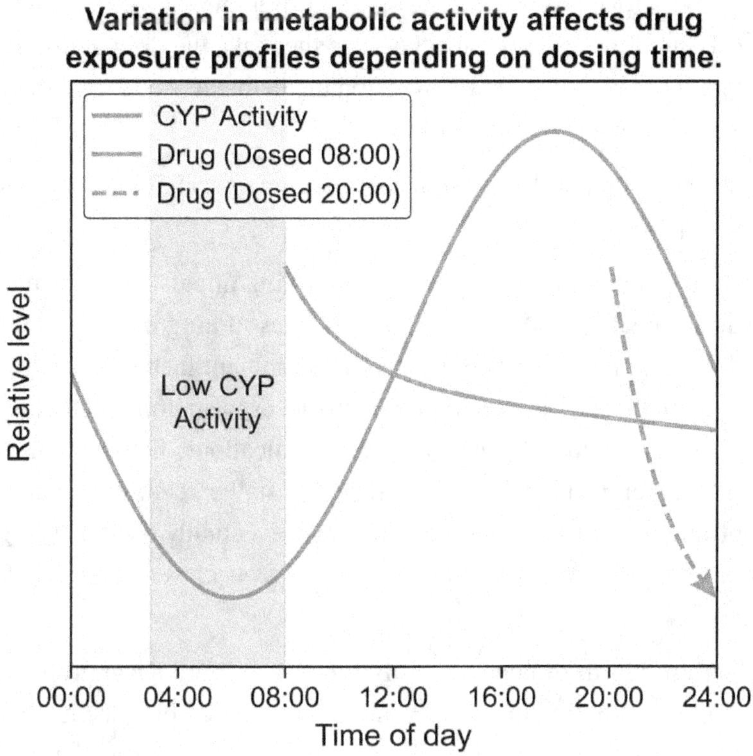

The point of the diagram is not the specific shape. It is the idea that "clearance" is not a constant, and that a clinician's mental model of a drug curve may be wrong by a clock shift. If metabolic enzyme activity peaks at one time, a dose taken just before that peak may be cleared faster, producing a lower and shorter exposure. A dose taken at a trough may linger, producing higher exposure and, potentially, more side effects. Whether that is good or bad depends on the drug, the target, and the patient's vulnerability window.

It also suggests a quiet reason why some drug trials are noisy. If time of dosing is not controlled or even recorded, it becomes an unmeasured variable that can inflate variability in both efficacy and toxicity.

Two participants receiving the same milligrams may effectively be receiving different exposures because they take the drug at different internal times. In the era of precision medicine, it is odd to sequence genomes while ignoring clocks.

The final practical lesson returns to a human-scale question: what does a patient do with this knowledge?

Not, in general, to change medication timing impulsively. For many drugs, especially those with long half-lives, timing may make little difference, and adherence is often the dominant factor. A perfectly timed pill that is forgotten half the time loses to an imperfectly timed pill that is reliably taken. For some medications, however, timing is already embedded in practice (evening statins, bedtime sedatives, morning diuretics), and for others it may eventually become a negotiated part of care, especially when side effects cluster at predictable hours.

What the fourth dimension really offers is a more accurate mental picture of what treatment is. A pill is not a static object with a fixed effect. It is a signal arriving into a body that is already oscillating. The liver reads that signal through enzymes and transporters that rise and fall like a tide, and the rest of the body feels the result as changing exposure. When timing is treated as a nuisance, the tide remains, and the patient is left to experience it as unpredictable luck. When timing is treated as part of pharmacology, the same tide becomes something medicine can finally begin to use.

Chapter 9

The Measured Life

As biological timekeepers age, their ticking grows quieter, fragmenting the cohesive rhythms of youth and altering how the body interacts with the solar day. We now attempt to recapture these fading signals through technology, trading the precise but cumbersome assays of the clinic for the algorithmic approximations of wrist-worn sensors. This quantification heralds a new era of ethical complexity, asking not just how we measure our internal time, but how society might exploit that knowledge.

9.1 The Damping Pendulum

A good clock does two things at once: it keeps time, and it *announces* time. The second part is easy to miss. You can own a perfectly accurate watch, but if its hands barely move you would still struggle to tell whether it is noon or midnight at a glance. Something similar happens in aging. The problem is not only that daily rhythms may drift earlier or later, but that the body's rhythmic signals often grow quieter. Peaks become less peaky; troughs less deep. Chronobiologists call this change in height *amplitude*. A rhythm can keep roughly the right period (about a day) and yet lose amplitude until it becomes hard for the body to use.

Amplitude matters because tissues do not just passively receive time.

They decide how strongly to respond. A strong circadian signal is like a clear conductor's beat: liver, muscle, pancreas, immune cells, and brain circuits can coordinate their entrances. A weak one is more like a conductor seen through fog: each section can still play, but the ensemble loses crispness. The result is not a single dramatic failure so much as a thousand small frays: lighter sleep that breaks apart, body temperature rhythms that flatten, appetite and digestion that become less predictably timed, and a reduced ability to absorb a late dinner or a disrupted night without paying a price the next day.

Amplitude is visible in one of the most familiar human circadian outputs: melatonin. Melatonin is a hormone produced by the pineal gland in dim light and darkness, with levels rising in the evening, peaking in the middle of the night, and falling toward morning. It is sometimes described as the body's "hormone of darkness." That phrase is not strictly accurate—melatonin is not sleep itself, and plenty of people can sleep with low melatonin—but it captures an important point: melatonin is one of the body's clearest nightly announcements that internal time has entered its biological night.

With aging, melatonin profiles often look less like a mountain and more like a hill. The timing of onset may shift for some individuals, but the striking visual change is frequently the height of the nocturnal peak. Even when older adults maintain a recognizable nightly rise, the peak can be lower and the curve more blunted. Part of the story is upstream: the central clock in the brain can send a weaker signal to the pineal gland. Part is downstream: the pineal and its pathways can change, and the body's responsiveness to melatonin can shift. In practice, what matters is the lived experience: a nightly hormone wave that used to feel like a tide becomes more like a gentle ripple.

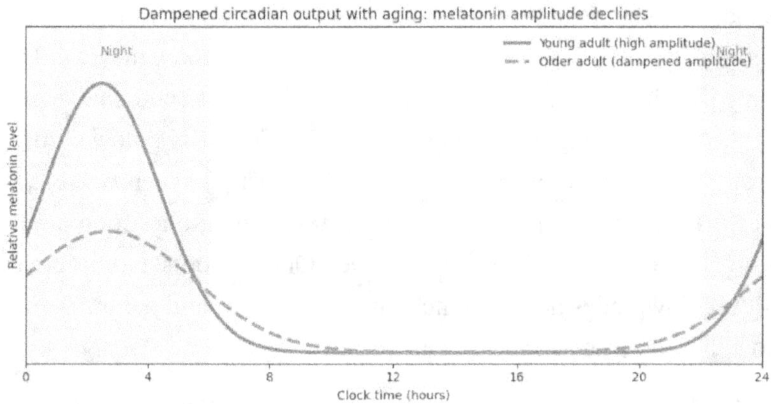

Once amplitude is on your mind, many everyday observations about aging look different. Consider the common complaint: "I fall asleep fine, but I wake at 3 or 4 a.m. and can't get back to sleep." It is tempting to interpret this as a simple phase shift, an internal clock that has moved earlier. Sometimes it is. But the same complaint can reflect a weaker signal rather than a strictly shifted one. Nighttime sleep depends on at least two forces working together: a sleep pressure that builds with time awake, and a circadian drive that makes sleep easier at night and wakefulness easier in the day. When the circadian contribution is strong, it helps protect the latter half of the night, even as sleep pressure is gradually "spent." When the circadian contribution is weaker, the second half of the night becomes fragile. Small disturbances—a creaking floorboard, a neighbor's car door, a pang of worry, a slightly warm bedroom, a bladder that no longer tolerates long intervals—are more likely to tip the system into wakefulness.

To see why aging so often attacks amplitude, it helps to replace the picture of the circadian clock as a single metronome with a picture of it as a *network*. At the center sits the suprachiasmatic nucleus (SCN), a small pair of structures in the hypothalamus. Each contains thousands of neurons, many of which are capable of generating rhythms on their own. If you isolate individual SCN neurons, they can keep time like tiny cellular watches. The surprise is that good timekeeping at the level of single cells is not enough. The SCN's power comes from its collective behavior: those cells must be synchronized, not perfectly but well enough that their peaks align into a strong population signal. Think of a stadium wave. One person standing up and sitting down does not announce much. Ten thousand people doing it together becomes unmistakable.

Aging appears to erode this collective synchrony. The SCN can lose neurons and synaptic connections; signaling molecules that help cells coordinate can change in abundance or effectiveness; electrical activity rhythms can become smaller in amplitude. Even if many individual neurons remain capable of oscillating, the network becomes less tightly coupled. In practical terms, the SCN starts to whisper where it once spoke. Downstream organs still have clocks, but they receive a less coherent broadcast. The result is not a full stop; it is a flattening.

Human evidence for this flattening is necessarily indirect, because you cannot easily sample the SCN of living people. Yet there are rare windows into the biology. In the 1990s, neuroanatomical work by Hofman and Swaab examined the SCN in postmortem human brains and reported age-related changes in the rhythmic pattern of vasopressin-expressing neurons, a key SCN output signal. The details are technical, but the headline is plain: a marker that looked strongly rhythmic in younger adults appeared reduced and disrupted in older adults. It is the kind of finding that brings the metaphor down to earth. "The clock weakens" is not only a complaint heard in bedrooms and nursing homes; it is also a pattern written into the tissue of the central pacemaker itself.

The other major contributor is light. Light is the dominant cue that aligns the SCN to the 24-hour day, and it does so through specialized retinal ganglion cells that are particularly sensitive to blue-enriched light. With age, the lens of the eye yellows and transmits less blue light; pupils become smaller; and some retinal function declines. Even without eye disease, the same lamp produces a different biological signal at age 70 than at age 20. This matters because the circadian system is not merely a clock; it is a clock that expects daily correction. When that correction is weaker, the system becomes easier to nudge off course and harder to stabilize. The body may still know the day is 24 hours long, but it hears the dawn less clearly.

A concrete historical reminder of how much circadian timing depends on light comes from long before modern chronobiology had its vocabulary. In 1729, the French astronomer Jean-Jacques d'Ortous de Mairan placed a mimosa plant in constant darkness and observed that its leaves continued to open and close on a daily schedule. This was a shock to the era's common sense: the plant seemed to carry its own day within it. The experiment is often cited as an early demonstration of endogenous rhythms. But it also contains an implicit warning. De Mairan did not claim the plant would thrive in permanent darkness. A rhythm can persist without the usual cues, but persistence is not the same as resilience. The circadian system evolved in dialogue with the environment. In aging, that dialogue can degrade: the cues arrive weakened, and the internal network that should respond to them may be less able to amplify and distribute them.

This is one reason why older adults can show a peculiar combination: a tendency toward earlier sleep timing *and* a tendency toward irregularity. Earlier timing is often called "phase advance" in the language of chronobiology. Irregularity is more subtle. It may show up as naps that move around the day, meals that drift, sleep that fragments, and activity that becomes more sensitive to social schedules, pain, medication effects, or caregiving routines. The clock is still there, but its grip on behavior loosens.

The body then faces a coordination problem. Peripheral organs—liver, gut, adipose tissue, skeletal muscle—contain molecular clocks built from feedback loops of gene expression. These local clocks are not puppets. They can be shifted by cues such as food timing, exercise, temperature, and hormones. In youth, the SCN's signal usually dominates, and the other cues often align with it: meals tend to occur during biological day, movement is concentrated in daylight, and sleep consolidates at night. In older age, if sleep becomes fragmented and eating times become more variable, those peripheral clocks may receive mixed messages. The liver may begin its metabolic day at one time, the pancreas at another, and muscle at another. The body can tolerate some misalignment. But when amplitude falls and synchrony weakens, tolerance shrinks.

A familiar example is the "late dinner problem." Many younger adults can eat a heavy meal at 9 or 10 p.m. without immediate catastrophe. They may not feel their best, but the system is robust. In older adults, the same meal can be followed by reflux, restless sleep, and a morning that feels foggier than it should. The immediate causes are multiple—slower gastric emptying, changes in muscle tone at the esophageal sphincter, medication effects—yet circadian amplitude adds an important layer. Metabolic pathways that are normally prepared for nighttime fasting may be less strongly engaged, and the signals that partition the day into "handle food now" versus "repair and maintain now" can be less decisive.

There is also a paradox worth addressing, because it causes confusion and sometimes unnecessary worry. People hear that aging reduces melatonin, and they conclude that insomnia in older age is a simple melatonin deficiency. That is rarely true. Older adults can have good sleep with low measured melatonin; younger adults can have insomnia with robust melatonin curves. Melatonin is better thought of as a marker and a messenger: it reflects the state of the central clock and helps communicate darkness to the body. A reduced melatonin amplitude can contribute to sleep fragility, but it is not identical to the experience of insomnia. The danger in collapsing these ideas is that it turns a network problem into a single-pill problem.

At the same time, it would be a mistake to treat amplitude as an abstract laboratory curiosity. In elder care settings, amplitude has practical consequences that can be heartbreaking. In hospitals and nursing homes, older adults may spend much of the day under dim indoor light and much of the night exposed to interruptions and illumination. The environment thereby provides the circadian system with a damaging message: days and nights are not very different. For a robust circadian network, the body can still infer the difference. For a damped network, the inference may fail. Nighttime wakefulness increases, daytime sleepiness increases, and the boundary between day and night blurs. When this blurring meets cognitive vulnerability, delirium becomes more likely. Families often experience this as personality change: a parent who was oriented and gentle becomes agitated at night, or confused about time, or unusually sleepy in the afternoon. It is easy to attribute this entirely to "the brain aging," but part of it is a brain trying to operate without a strong temporal scaffold.

The phrase "damping pendulum" captures this well because it suggests a system that is still moving but losing energy with each swing. A pendulum in a vacuum keeps a steady amplitude. A pendulum in air gradually slows and swings less widely. The circadian system is not a pendulum, but it shares a key feature: it is a rhythmic system that depends on both internal drive and external reinforcement. Aging can increase the "friction" inside the system (through neural and molecular changes) and reduce the "push" from the outside (through reduced light input and more irregular daily schedules). The swing becomes smaller.

One can even see the idea of damping in how older adults respond to time cues. A strong circadian system can be shifted by a carefully timed light exposure, but it also resists random nudges. A weaker one may drift more readily when routines change. Retirement is a classic trigger. For decades, an alarm clock, commute, and workplace schedule impose a social day that often fortifies the biological one. After retirement, those constraints loosen. For some, this is liberating and sleep improves. For others, the day becomes less structured, light exposure may decline, and rhythms weaken further. The person experiences this as "I don't sleep like I used to," but the deeper story is "my days no longer teach my body what day is."

It is also worth being honest about variability. Not everyone's circadian amplitude declines in the same way or at the same rate. Genetics matters, health status matters, and so do habits that have little glamour but large influence: daily time outdoors, consistent meal timing, regular physical activity, and preserving darkness at night. Neurodegenerative diseases can accelerate the damping dramatically, but even among healthy older adults the range is wide. Some seventy-year-olds maintain stable sleep and strong daily rhythms; some fifty-year-olds already show pronounced fragmentation. Aging is not a single track; it is a landscape.

Yet the common direction remains: the circadian network's outputs tend to flatten. When the body cannot clearly distinguish biological day from biological night, many downstream systems adopt compromise states: not fully in day-mode, not fully in night-mode. That compromise may be tolerable for a while, but it is rarely optimal. Repair processes that prefer consolidated night can be interrupted. Daytime alertness can become patchy. Metabolic flexibility can decrease. The person may not describe this in circadian terms; they may simply say, “I feel less steady.”

There is a quiet psychological effect here, too. A strong circadian rhythm gives a sense of predictable selfhood: morning energy, afternoon steadiness, evening winding down. When rhythms damp, the day can feel less legible. People sometimes interpret this as a failure of willpower—“I should be able to sleep through the night” or “I should not need a nap”—when it is often a change in internal signaling. The body is not being stubborn. It is being less clearly instructed.

What restores amplitude is not a mystery in principle. The circadian system responds to contrast: bright days, dark nights, regular timing of food and activity, and enough quiet to let sleep consolidate. The difficulty is that aging, illness, and modern living often reduce contrast precisely when the system needs it most. Older eyes collect less light; older bodies are more sensitive to disrupted nights; caregiving schedules and medications can fragment days; safe neighborhoods and mobility limitations can keep people indoors. Damping is therefore both biology and biography.

And that is perhaps the most human part of the science. A circadian rhythm is not only a curve on a graph or a gene loop in a cell. It is a daily pattern that is continually negotiated between an organism and its world. In later life, the negotiation becomes harder, not because the clock disappears, but because its voice softens. The task becomes less about perfect timing and more about giving that soft voice something worth synchronizing to: a morning that is unmistakably bright, a night that is reliably dark, and a day that still has shape enough for the body to recognize itself within it.

9.2 Signals in the Noise

When people say they want to "know their circadian rhythm," they usually mean one of two things. Some want to know *phase*: where they are in their internal day right now, the way you might want to know whether your body thinks it is late afternoon or midnight. Others want to know *strength*: whether their rhythms are crisp or blurred, the difference between a clear radio station and a hissy one. The two questions are related, but they are not the same, and modern technology has encouraged the hopeful belief that both can be answered by a ring, a watch, or a phone on the bedside table.

There is a reason clinicians and chronobiologists hesitate before promising that kind of certainty. The body keeps time in many places at once, and most of those clocks are not directly visible. What a wearable can see is behavior and a few surface traces of physiology: movement, skin temperature, pulse waves, sometimes a crude measure of light. From those traces it tries to infer internal time, the way one might infer the schedule of an unseen train station by watching the flow of taxis outside. The inference can be useful. It can also be wrong in systematic, convincing ways.

The clinical gold standard for measuring circadian phase in humans is the *dim light melatonin onset*, abbreviated DLMO. The name sounds more complicated than the idea. Melatonin levels are low during the day and rise in the evening, but light suppresses melatonin. If you want to measure *when* the body begins its biological night, you have to take light out of the equation. So you place a person in dim light for several hours before bedtime and collect saliva samples repeatedly, often every 30 minutes, to detect the moment melatonin crosses a threshold from "daytime low" to "nighttime rising." That crossing point is DLMO. It is not a perfect marker, but it is a remarkably informative one. It tends to be stable within individuals under stable routines, it shifts in predictable ways with light exposure, and it anchors many research findings about jet lag, shift work, delayed sleep-wake phase disorder, and seasonal changes.

DLMO is also, for most people, profoundly impractical. It is the kind of measurement that fits a sleep lab, not a Tuesday evening. The "dim" in dim light is not poetic. It means truly low illumination, often tens of lux or less, the sort of lighting you might associate with a restaurant trying to be romantic or a hallway nightlight. A bright kitchen, an overhead bathroom lamp, a phone screen at full blast: any of these can suppress melatonin and push the rise later, or flatten it enough to make the threshold crossing ambiguous. Even the act of sampling requires discipline. You cannot eat or brush your teeth right before a saliva sample; you must label tubes correctly; you must keep the timing exact. The measurement asks for a kind of gentle captivity: sit in dimness, do not scroll, do not wander into a bright room, keep collecting samples, and wait for your own biology to reveal itself.

That is what makes DLMO such a useful bridge between concept and lived experience. It is the opposite of a vague wellness claim. It is a protocol that can fail if you do not follow it. It also embodies an uncomfortable truth about measurement: the more precisely you want to know something, the more you must control the world around it.

The difficulty is that most of modern life runs on the opposite principle. We want measurements that are continuous, effortless, and cheap, gathered in the background while we do everything else. That desire has produced a booming ecosystem of consumer sleep and circadian technologies. The basic ingredients are familiar: an accelerometer that detects motion, a heart-rate sensor that detects pulse via light reflected from the skin, sometimes a skin temperature sensor, sometimes an estimate of blood oxygen, occasionally a light sensor, and an algorithm trained to translate those streams into tidy outputs: “sleep stages,” “recovery,” “readiness,” “chronotype,” “sleep debt,” and increasingly, an estimate of circadian phase.

It is tempting to treat those outputs as if they were direct readings of internal time. But they are not. They are inferences, and they are constrained by two problems that sound technical but are deeply human: *proxy* and *noise*.

A proxy is a stand-in. Movement is a proxy for sleep, because sleeping people tend to move less. Heart rate patterns are a proxy for certain sleep features, because autonomic tone changes across the night. Skin temperature is a proxy for circadian phase, because peripheral temperature rhythms relate to vasodilation and internal timing. Proxies are unavoidable when direct measurement is hard. The question is how far the proxy can take you before it begins to tell a story you want to hear rather than the one your body is actually living.

Noise is everything else: random fluctuations, sensor errors, algorithmic assumptions, and the sheer complexity of human nights. A watch can slip on the wrist. A photoplethysmography heart-rate signal can degrade with tattoos, darker skin tones, motion, cold hands, or poor contact. A temperature sensor measures skin, not core. Even if the sensors were flawless, human sleep is not a clean laboratory phenomenon. A person may lie still while awake, especially if anxious or trying to “be good” for the tracker. Another may move in sleep. Alcohol may reduce movement but fragment physiology. Medications can alter heart rate and thermoregulation. A fever, a late meal, or a stressful day can distort the patterns that algorithms expect.

In the clinic, the temptation is to push back against this mess by adding more wires and more control. The most definitive sleep test is polysomnography: brain waves (EEG), eye movements (EOG), muscle tone (EMG), respiratory measures, and more. It is expensive and intrusive, and it is not intended for nightly use. Actigraphy, a clinical method based primarily on movement, sits somewhere between the lab and the bedroom. It is useful for measuring sleep timing and duration over many nights, especially in circadian rhythm sleep-wake disorders, insomnia, and in populations where lab studies are hard. But actigraphy is not the same as a consumer wearable, and even actigraphy has limits. Movement can estimate when someone is likely asleep, not precisely which stage they are in. Efforts to extract sleep stages from actigraphy-like signals exist, but validation is uneven and performance varies.

The consumer world tends to flatten these distinctions. Many devices present "deep sleep" and "REM sleep" as if they were measured with the same authority as in a sleep laboratory. This can lead to a peculiar modern anxiety: people begin to distrust their own experience. They wake feeling fine but the app says their deep sleep was low; they feel tired but the app congratulates them on eight hours. Some become preoccupied with optimizing scores, adjusting routines not for comfort or health but to appease the device. Sleep clinicians have a name for this: *orthosomnia*, the pursuit of perfect sleep metrics that paradoxically worsens sleep by making it a performance.

Orthosomnia is not just a quirky psychological trend. It exposes something fundamental about measurement. Data does not simply describe; it also persuades. A number displayed with decimals carries an aura of objectivity, even when it is the output of a model built on averages and assumptions. When a device tells you that you had 1 hour and 12 minutes of "deep sleep," it implies that there is a stable, measurable thing called deep sleep that your wrist can detect and that your body can bank like money. The truth is messier. Sleep stages are defined by brain activity and other physiological signals. A wearable can guess based on patterns correlated with stages in validation studies, but correlation is not identity.

Circadian phase estimation is even more challenging, because phase is not a single thing you can point to on the body. It is an alignment across many rhythms. DLMO became a gold standard partly because melatonin is a relatively clean output signal under controlled conditions: it rises in the evening in a fairly orderly manner. In everyday life, melatonin is hidden behind light exposure, and light exposure is not just a number on a sensor. It is where the light lands in the retina, how long it lasts, what spectrum it contains, and what your eyes and brain do with it. A wrist light sensor, even when present, measures light on the wrist, which can differ dramatically from light in the eyes. A sleeve, a blanket, the angle of the arm, a lamp behind you: all can break the link between what the watch measures and what the circadian system receives.

This is why wearable-based circadian phase prediction can be surprisingly uncertain, even when it sounds confident. Models that incorporate multiple streams (activity, light exposure, sometimes heart rate) can outperform crude guesses based solely on bedtime. Yet in real-world validation, especially in shift workers whose schedules and light exposure are irregular, prediction errors on the order of hours are common. An error of two or three hours is not trivial. It is the difference between being in your biological evening and being in your biological night, the difference between light that gently reinforces your rhythm and light that powerfully shifts it, the difference between a meal your metabolism is prepared for and one it will struggle to handle.

The most instructive failures happen when a person's behavior decouples from their internal clock. Consider a night shift nurse who sleeps from 9 a.m. to 2 p.m. after work, then naps from 7 p.m. to 9 p.m., then works overnight. Movement-based systems can easily interpret this as chaotic sleep. A consumer algorithm may "correct" the chaos by forcing it into a normal template: long nighttime sleep, short daytime naps. But the nurse's internal circadian phase might be in the slow process of shifting, partially anchored by morning daylight during the commute home, partially shifted by bright hospital lighting at night, partially shaped by days off. What looks like messy behavior may be the outward expression of a circadian network under competing demands. A simplistic algorithm can mistake adaptation for pathology, or vice versa.

There is a historical parallel here that is almost too perfect. When telegraphy spread in the nineteenth century, timekeeping became a crisis. Before railways and telegraphs, towns kept local time based on the sun. Noon was when the sun was highest. But once trains began crossing regions quickly, and messages began leaping across distances instantly, "noon" became a local idiosyncrasy that could cause collisions and missed connections. The solution was standard time zones and synchronized clocks. Measurement created a new kind of order, but it also created a new kind of anxiety: if time is now a number coordinated across society, being wrong by even a few minutes matters.

Wearables have done something similar to sleep and circadian life. They have created a culture in which internal states are expected to be quantifiable and comparable. A person can now wake up and find themselves labeled as "behind" or "recovered," "on schedule" or "misaligned." The advantage is obvious. For many, these devices reveal patterns that were previously invisible: a late coffee that reliably fragments sleep, a weekend schedule that produces Monday fatigue, an alcohol night that looks restorative by hours but not by physiology. They can prompt healthier habits and encourage regularity.

The danger is the illusion of precision. In medicine, a measurement earns authority not because it can be displayed, but because it has been validated for a specific use. DLMO is validated for circadian phase, under controlled lighting, with assays sensitive enough to detect low levels, and with sampling frequent enough to detect a threshold crossing. Even then, there are subtleties: where you set the threshold matters; some people secrete little melatonin and become difficult to classify; and small protocol violations can shift the result.

Attempts to make DLMO easier have moved it from the lab to the home. This sounds like a simple improvement: send people home with saliva tubes and instructions, and they can sit in their own living room rather than a laboratory. In practice, home DLMO can work well when compliance is good and when the protocol is carefully designed, but it comes with predictable failure modes. People forget a sample or collect it at the wrong time. They turn on brighter lights without realizing the impact. They swap the order of tubes. Some individuals simply do not produce a strong enough melatonin rise for a threshold-based onset to be determined reliably, even with good sampling. Studies comparing home and lab DLMO often find overall correlation with typical differences measured in tens of minutes, but the outliers are the ones that matter clinically: the person whose DLMO appears "normal" at home because the evening was brighter than intended, or the person whose onset is indeterminate because of low secretion. Adding objective compliance measures—ways to track light exposure and sampling times—can tighten the agreement substantially, but that is precisely the point: the accuracy comes from controlling and monitoring the conditions, not from the mere fact of being at home.

Consumer wearables take a different approach. Instead of measuring melatonin, they attempt to infer circadian timing from patterns over many days. This can be genuinely informative at the population level and can offer a rough estimate for an individual. If you want to know whether your schedule is stable, whether your mid-sleep point is drifting later, whether you are getting bright light in the morning or spending most of the day in dim indoor illumination, a wearable can provide clues. What it cannot yet do reliably for most people is provide a DLMO-equivalent timestamp with clinical confidence. The device may show you a confident number because confidence is part of product design. Biology does not always cooperate.

The mismatch between what people want and what devices can deliver is nowhere clearer than in the language of “sleep stages” on consumer dashboards. The stages themselves are real, but the wearable’s access to them is indirect. It uses movement and pulse patterns to estimate the probability of a stage at a given moment, then stitches those probabilities into a narrative of your night. If the wearable is wrong, it is often wrong in plausible ways. It produces an output that looks like sleep, because the algorithm is optimized to produce outputs that resemble sleep. A person can lie awake, perfectly still, and be scored as asleep. Another can toss and turn in genuine sleep and be scored as wake. A night of fragmented sleep can be smoothed into a seemingly orderly architecture. That smoothing is soothing to look at, but it is not always truthful.

Still, it would be a mistake to dismiss the entire consumer world as noise. Some of the most valuable circadian insights come precisely from long-term, messy measurements. The circadian system is not only a clock; it is a pattern of regularities and deviations. A laboratory snapshot can miss the lived rhythm. A wearable that tracks a month of bedtimes can reveal social jet lag more clearly than a single night in a lab. A week of activity and light exposure can show whether someone spends their mornings in darkness and their evenings under bright screens, a pattern that almost invites delayed circadian timing. For older adults with damped rhythms, continuous tracking can reveal how often nights are interrupted and how strongly daytime activity consolidates, information that may guide practical changes in routine.

The key is to treat wearable data as a *hint* rather than a verdict. A hint is something you test against experience and context. If the device suggests your sleep timing is irregular, you can ask: is that because of my schedule, or because I am lying awake worrying? If it suggests you are not getting morning light, you can check your routine: do you actually go outside before noon? If it suggests your resting heart rate rises after alcohol, you can notice whether you also wake earlier. In other words, the data becomes part of a conversation with your body, not a replacement for it.

One practical way to hold onto this perspective is to remember what each method truly measures. DLMO measures melatonin rise under controlled dim light, and is best for pinpointing circadian phase. Polysomnography measures brain and body signals directly, and is best for diagnosing sleep disorders and characterizing sleep architecture. Clinical actigraphy measures movement over time, and is best for sleep timing, duration, and regularity across days and weeks. Consumer wearables measure similar signals to actigraphy plus heart-rate-related proxies, and are best for personal pattern recognition, habit change, and rough tracking, with accuracy that varies by device, person, and situation.

The noise becomes most dangerous when it is mistaken for signal, and the signal becomes most valuable when it is used to improve life rather than to grade it. A person who learns, through a watch, that their "best" sleep happens when they keep weekend wake times closer to weekdays has learned something useful, even if the exact minutes of "deep sleep" are uncertain. A person who becomes afraid of going to bed because the watch might report a poor score has learned something harmful, even if the sensors are functioning perfectly.

The deeper issue is not whether consumer wearables are good or bad. It is that circadian biology has become legible to the public in a way that few internal systems have. Blood pressure cuffs and glucose meters made certain aspects of cardiovascular and metabolic health measurable at home, and they also changed behavior and self-conception. Wearables are doing that for sleep and timekeeping. The challenge is to preserve the humility that good measurement demands: the recognition that a number can be helpful without being final, and that the body's internal clocks are not a single meter to be read but a distributed network to be interpreted.

A clean signal is rare in biology. Most of the time, the truth arrives as a pattern that must be coaxed out of mess. The promise of modern circadian measurement is real, but it is not the promise of perfect control. It is the promise of better questions asked more often, in the places where people actually live: bedrooms with streetlights outside, kitchens with late dinners, night shifts with fluorescent glare, and mornings that begin, for many, with the small, bright rectangle of a phone.

9.3 Bio-Data and the Panopticon

A wristwatch used to be a private object. It told you the time, and perhaps, if you were the sort of person who enjoyed such things, it told you something about yourself: punctual, disorganized, careful with money, careless with money. The modern wristwatch is more ambitious. It does not simply display a clock; it tries to *be* a clock for your body, translating sleep, activity, light exposure, heart rhythms, and temperature into an annotated portrait of your days and nights. Once that portrait exists, it begins to behave like any other valuable data: it can be stored, copied, sold, subpoenaed, demanded, or quietly repurposed.

The unease people feel about this is not paranoia. It is a recognition that circadian data is peculiarly intimate. A credit card statement may reveal what you bought; a location trail may reveal where you went.

Circadian traces can reveal when you are most vulnerable, when you are cognitively sharp, when you are likely asleep, when you are likely exhausted, when you might be depressed, when you might be sick, when you might be working two jobs, when you might be caring for an infant, when you might be drinking more than you tell your doctor, when you might be taking medication regularly or irregularly. Even when those inferences are probabilistic and sometimes wrong, they can still be actionable in the hands of someone with power over you.

The image that often comes to mind is the panopticon: the circular prison design proposed by Jeremy Bentham in the late eighteenth century, in which a central guard tower could, in principle, observe any prisoner at any time. Bentham's architectural idea became, in later political theory, a metaphor for modern surveillance: the most effective observation is not constant watching but the *possibility* of watching, which changes behavior even when no one is looking. The connection to circadian measurement is not that employers or insurers are literally watching your sleep in real time. It is that the knowledge that your rhythms are measurable, recordable, and transferable can alter how you live in your own body.

A striking feature of circadian and sleep data is that it slips easily across the boundary between *health* and *performance*. A clinical measurement like DLMO aims to identify circadian phase for medical and research purposes, under controlled conditions, with known limitations. A consumer "readiness" score, by contrast, looks like health but behaves like productivity. It invites a question that is both seductive and dangerous: if a device can identify your best hours, why not schedule you accordingly?

In medicine, time-aware care can be genuinely beneficial. The logic is simple enough for a dinner party: drugs do not act on a static body. The liver's enzymes, the immune system's responsiveness, airway caliber, blood pressure, pain sensitivity, and clotting risk all change across the day. For some conditions, treatment timing can matter. This is the promise of chronotherapy: matching intervention to biological time. At its best, it is an act of respect for circadian networks rather than a demand that patients live like machines.

But the same logic can be turned outward. If biological time can be measured, it can also be used to rank, sort, and pressure. The future that worries people is not merely that someone might see their data, but that someone might say: "Your biological night begins early; therefore you should always work mornings," or "Your sleep looks fragmented; therefore you are a safety risk," or "Your rhythms look delayed; therefore you are less disciplined." These claims can be made with the confident tone of science even when the underlying measurements are noisy and the inferences are shaky.

There is a name for this possibility: *chronodiscrimination*. The word is new, but the underlying pattern is old. Employers have always judged workers by punctuality, alertness, and the ability to tolerate inconvenient schedules. What changes with wearables is that judgments can be anchored to data that looks objective, can be scaled across an organization, and can be interpreted without empathy. A manager does not need to understand the difference between sleep pressure and circadian phase, or the limits of actigraphy, to be persuaded by a dashboard.

One can imagine a well-meaning version first. A hospital wants to reduce medical errors on night shift. It offers staff an optional sleep and circadian program, with bright-light exposure strategies, protected recovery sleep, and better shift rotation. Participation is voluntary, data stays private, and workers feel helped rather than evaluated. This is plausible and, in some places, already partially real.

Now imagine the program becomes "standard." A worker who opts out is seen as not committed to safety. Or the data is used to justify staffing decisions: those with "poor recovery" are denied overtime; those with "better readiness" are assigned high-responsibility tasks; those with "early chronotype" are quietly pushed off late shifts, even if late shifts are higher paid. The program has shifted from health support to labor sorting, and it has done so without a single villain. That is often how surveillance becomes normal: not by a dramatic overreach, but by a series of reasonable-sounding steps.

The ethical tension is sharpened by a fact already visible in aging: circadian amplitude can weaken, rhythms can fragment, and resilience can fall. If circadian metrics become part of employment screening, older workers could be disadvantaged not only by overt ageism but by a biologized metric that amplifies existing inequities. The same risk applies to people with chronic illness, caregivers, new parents, and shift workers whose rhythms are, by definition, under stress. A system that claims to be optimizing performance can end up penalizing those whose lives are least compatible with idealized regularity.

Privacy debates often begin with ownership: who owns the data? Yet ownership is only the surface. The deeper issue is the set of properties that make digital health data uniquely slippery.

- *Invisibility*. People often do not know what is being collected, when, and by whom. A watch collects continuously. An app may share silently. A setting may change after an update. Even a careful user can be surprised.
- *Inaccuracy*. Wearable-derived sleep staging and circadian inference are proxies. They can be useful, but they can also be wrong in convincing ways. When the data is wrong, it does not simply fail to help; it can actively harm if decisions are made on its basis.
- *Immortality*. A bad night is usually forgotten. A logged bad

night can persist for years. A period of insomnia during a divorce can sit in a database long after life has stabilized, waiting to be reinterpreted.

- *Marketability*. Digital traces are not merely stored; they are assets. They can be bundled, analyzed, and monetized in ways that are opaque to the person whose body generated them.
- *Identifiability*. Even when data is stripped of obvious identifiers, patterns of sleep and movement can be distinctive, especially when combined with other datasets. A supposedly anonymous circadian profile may not remain anonymous for long.

These properties matter because, in the United States, health privacy protections are uneven. Many people assume that anything "health-related" is protected by HIPAA, the Health Insurance Portability and Accountability Act. HIPAA is important, but it is narrower than the public imagination. It applies to certain covered entities (healthcare providers, health plans, and their business associates) under specific conditions. A consumer sleep app or wearable company may fall outside HIPAA's scope while still collecting extremely sensitive information.

That gap has attracted regulatory attention from the Federal Trade Commission (FTC), which operates through consumer protection and competition law rather than medical privacy law. The FTC's Health Breach Notification Rule (first issued in 2009) is aimed at vendors of "personal health records" not covered by HIPAA, requiring notifications to consumers (and sometimes the FTC and the media) when there is a breach of unsecured identifiable health information. In April 2024, the FTC finalized updates intended to clarify and modernize this rule, explicitly addressing the contemporary world of health apps and connected devices and tightening requirements around breach notification content and timing. The thrust is plain: if you collect health data outside HIPAA, you do not get a free pass.

Enforcement actions can make these abstractions feel real. In 2023, the FTC took action against GoodRx, alleging that the company shared sensitive health information with advertising platforms despite promises to users. The remedies emphasized exactly the things people worry are impossible once data has spread: directing third parties to delete shared data, imposing retention limits, and requiring a comprehensive privacy program. One can debate how fully such remedies can unwind harm, but the case established a precedent: health-related data flows used for advertising can trigger serious consequences, and "we are not a hospital" is not a moral defense.

The corporate side of the story adds another layer: even when there is no breach and no advertising scandal, people can lose control through mundane platform governance. Accounts merge. Terms change. Services sunset. A person who has tracked sleep for a decade may discover that their ability to access, export, or keep that history depends on agreeing to a migration they did not request. The Fitbit-to-Google account transition is a vivid example because it comes with a clock attached: reports in 2025 described a firm deadline of February 2, 2026, after which users who have not migrated risk losing access to the Fitbit service, and historical data may not be retained beyond that date except as needed for legal compliance. Whatever one thinks of the business reasons, the human message is unmistakable: the record of your nights is not only about your body; it is also about the policies of a company.

This is where the "panopticon" metaphor becomes most useful, not as melodrama but as a description of asymmetry. The individual generates the data through living. The platform stores it. The platform sets the interface through which the individual can see themselves. The platform can change the interface, adjust what is shown, introduce new metrics, and decide what counts as "improvement." A person can resist by deleting an app, but deletion is not always total, and it comes with a cost: you lose the record that, in the first place, helped you notice patterns and advocate for yourself.

The workplace is not the only arena in which circadian data can become coercive. Insurance is an obvious candidate. If premiums are adjusted based on activity, why not based on sleep? The argument can be framed as fairness: if better sleep reduces health risk, why should those who "choose" poor sleep be subsidized? The problem is that sleep is not simply a choice. It is constrained by work schedules, neighborhood noise, caregiving, mental health, medication, chronic pain, and the damped amplitude of aging physiology. A system that prices sleep as if it were a leisure activity risks becoming a system that taxes vulnerability.

Clinical settings introduce a different kind of coercion: the quiet pressure to share data to be taken seriously. A patient arrives with fatigue, brain fog, or insomnia. They bring months of wearable graphs. Some clinicians welcome it; others are skeptical. The patient may feel they must produce data to justify their suffering. The wearable becomes a passport into medical attention. Yet the data may be misleading, and the patient may be punished for trusting it. The result can be a perverse loop: mistrust of one's own experience, dependence on imperfect numbers, and a clinical conversation distorted by metrics that were never designed for diagnosis.

A more hopeful path exists, but it requires clarity about boundaries. Circadian measurement can support health without becoming surveillance if several conditions hold. Consent must be genuine, not bundled into employment or access to care. Data minimization must be real, not rhetorical: collect what is needed for the stated purpose, not everything that can be collected. Retention must be limited, because a life is allowed to change. Access must be layered, because the person is not merely a data source but the primary stakeholder. And interpretation must remain humble, because circadian proxies are not the same as circadian truth.

Perhaps the most practical ethical test is this: does measurement increase a person's agency, or does it shift agency elsewhere? If a shift worker uses light and sleep timing strategies to reduce misalignment, that is agency. If a platform uses readiness scores to pressure longer hours, that is extraction. If an older adult uses a rhythm record to negotiate better nighttime lighting in a care facility, that is agency. If a facility uses the same record to justify sedating someone at night for institutional convenience, that is coercion.

The body's clocks evolved to coordinate tissues, not to produce performance reports. The more precisely circadian life can be measured, the more urgently society must decide what counts as care and what counts as control, because the difference is not in the sensors. It is in who gets to interpret the signal, and what they are allowed to do with it.

www.ingramcontent.com/pod-product-compliance
Lightning Source LLC
LaVergne TN
LVHW022205200926
848481LV00006B/1852
* 9 7 9 8 8 9 6 6 5 2 8 5 4 *